CHARLES DENTON ABEL

RUDIMENTARY

AND

ELEMENTARY PRINCIPLES

OF THE

CONSTRUCTION AND ON THE WORKING OF

MACHINERY

ILLUSTRATED

BY NUMEROUS EXAMPLES OF MODERN MACHINERY FOR DIFFERENT BRANCHES OF MANUFACTURE

Elibron Classics
www.elibron.com

Elibron Classics series.

ISBN 1-4212-2179-9 (paperback)
ISBN 1-4212-2178-0 (hardcover)

This Elibron Classics Replica Edition is an unabridged facsimile of the edition published in 1860 by John Weale, London .

RUDIMENTARY

AND

ELEMENTARY PRINCIPLES

OF THE

CONSTRUCTION AND ON THE WORKING OF

MACHINERY,

ILLUSTRATED

BY NUMEROUS EXAMPLES OF MODERN MACHINERY FOR DIFFERENT BRANCHES OF MANUFACTURE.

BY

C. D. ABEL, C.E.

LONDON:
JOHN WEALE, 59, HIGH HOLBORN.
1860.

LONDON:
BRADBURY AND EVANS, PRINTERS, WHITEFRIARS.

PREFACE.

In the following pages it has been endeavoured to give to the student a general insight into the principles by which the working of machinery is regulated, and upon which depends its geometrical and mechanical construction, firstly, by investigating the physiology of motive power in its different modifications, its various sources and modes of action.

Secondly, by inquiring into the manner in which the motion of machinery is influenced, by the motive power acting upon it, as also the resistance to be overcome by it, being varied both as regards their intensity and quantity, and showing what principles of construction are the best adapted for certain cases.

Thirdly, by enumerating and considering the various circumstances which induce a diminution in the effectiveness of the motive power exerted upon a machine, and pointing out the methods that may be adopted for computing the same.

Fourthly, by describing the various contrivances that

have been adopted for so regulating the velocity with which a machine works as to maintain that degree at which the machine acts to the greatest advantage.

Fifthly, by investigating in general, and in detail, the principles of construction of machinery; and,

Lastly, by illustrating the practical application of the latter in describing a variety of machines involved in some of the most important branches of manufacture.

To have dealt with these various subjects more than superficially, would have expanded this treatise to a size far exceeding the limits within which the works designed for this series have to be kept, and on this account also the author has been obliged to omit entering at all into the description of that most important of all machines, the steam-engine, on which, however, there already exists a treatise in this series; also the intention of including the construction of water-wheels and turbines in this work had to be abandoned, the accompanying atlas of plates having already exceeded the limits first assigned to it. With regard to the atlas, the author begs to state that he is indebted to Sir William Armstrong for the drawing of hydraulic machinery; to Mr. Anderson for those of the Minié bullet machinery; to Mr. Cowper for that of the Crystal Palace Waterworks pumping-engine; and to Mr. Horne for the drawing of sawing-machinery: to which gentlemen he begs to express his best thanks.

It has been the endeavour of the author to omit the use of mathematical manipulation as much as circumstances would permit of, so as to render the subject as intelligible as possible to those not versed in that science. He felt himself justified in employing the co-ordinate system for the graphical illustration of the action of forces, as it is easily understood, while affording the clearest possible representation of static and dynamic laws.

CONTENTS.

CHAPTER II.

OF THE MANNER IN WHICH THE MOTION OF MACHINERY IS INFLUENCED AND MODIFIED BY DYNAMICAL LAWS.

CHAPTER III.

OF THE USEFUL AND DETRIMENTAL EFFECTS OCCURRING IN THE ACTION OF MACHINERY IN MOTION.

CHAPTER IV.

OF THE CONTRIVANCES FOR REGULATING THE MOTION OF MACHINES.

CHAPTER V.

ON THE GENERAL PRINCIPLES OF CONSTRUCTION OF MACHINERY.

CHAPTER VI.

OF THE SPECIAL CONSTRUCTION OF MECHANICAL COMBINATIONS.

CHAPTER VII.

OF VARIOUS DESCRIPTIONS OF MACHINES, CONSTRUCTED FOR DIFFERENT PROCESSES OF MANUFACTURE.

SECTION I.

SECTION II.

SECTION III.

SECTION IV.

SECTION V.

CHAPTER VIII.

OF PUMPS AND PUMPING ENGINES.

SECTION I.

SECTION II.

SECTION III.

CHAPTER IX.

OF HYDRAULIC CRANES.

LIST OF PLATES.

PLATE I.

PLATE II.

PLATE III.

PLATE IV.

PLATE V.

PLATE VI.

PLATE VII.

MACHINERY FOR THE MANUFACTURE OF IRON.

PLATE VIII.

MACHINERY FOR THE MANUFACTURE OF IRON.

PLATE IX.

MACHINERY FOR THE MANUFACTURE OF COTTON.

PLATE X.

MACHINERY FOR THE MANUFACTURE OF COTTON.

PLATE XI.

MACHINERY FOR THE MANUFACTURE OF COTTON.

PLATE XII.

MACHINERY FOR THE MANUFACTURE OF PAPER.

PLATE XIII.

MACHINERY FOR THE MANUFACTURE OF CONICAL BULLETS.

PLATE XIV.

MACHINERY FOR THE MANUFACTURE OF CONICAL BULLETS.

PLATE XV.

MACHINERY FOR THE CONVERSION OF TIMBER.

PLATE XVI.

PUMPS AND PUMPING-ENGINES.

PLATE XVII.

HYDRAULIC CRANE.

ELEMENTARY PRINCIPLES OF MACHINERY.

INTRODUCTION.

1. FROM the remotest ages, man was wont to explore the hidden treasures of nature in search of such things as could contribute to his bodily comfort. In his uncivilised state, his intellectual faculties were not sufficiently developed for him to appreciate more than such objects as immediately presented themselves in a palpable state of utility. He was content with the fruits and animals of the forest for his food, their skins for his clothing, and with rough wigwams for shelter. As civilisation gradually dawned upon him, his intellect expanded and, in proportion, his wants increased. His attention, through the guidance of Providence, was drawn to objects presented to him by nature, which at a first glance might seem useless to him, yet, on further investigation, he would find only to need certain transformations to satisfy his wants. These *necessary transformations* originated that branch of industry termed manufacture, which; becoming daily more and more intricate in its processes, at length compelled man to invent some contrivance by which a more certain and rapid result might be arrived at than could be obtained by mere manual labour.

This contrivance was no less than the *machine* in its widest sense; that is, a mediator between *power* and the substance to be transformed. The next step onward was caused by the inadequacy of human and animal power to perform, even with the aid of the machine, the duties assigned to them; then it was that man recognised the sources of power lying dormant in nature; and he contrived his machine so that it should be acted upon by and turn to useful account, these natural forces; and thus originated that class of machines termed *engines*.

2. A *mechanical process* consists of four elements, namely: *a*, the motive power; *b*, the receiver and transmitter of the same, or engine; *c*, the working machine or tool; and *d*, the substance to be transformed. In any mechanical process of magnitude, we shall be able further to classify the machinery employed into three distinct parts, namely: *a*, the engine or receiver of power; *b*, the transmitter of power, generally consisting of lines of shafting, gearing, &c., which convey the power to *c*, the working machines. On the other hand, in some systems of machinery we shall find the receiver of the motive power in direct connection with the working machine, forming in fact part of the same. But in all cases we shall easily distinguish this feature, though differing in its construction according to the nature of the motive power acting upon it. Thus, in the steam-engine it is the cylinder and piston; in the mill driven by water-power, the water-wheel or turbine; in the windmill, the sails; in machines worked by manual labour, the winch, &c. In general we may say that the duty of the receiver is to control the action of the motive power, to force it to convert its irregular and indefinite motion into a regular, generally a rotatory, or rectilinear alternate motion. The duty of the transmitter is to convey, and to increase or decrease the speed of the power, likewise in a perfectly definite and unchangeable manner; that of the working machine to change the motion to the description required,

and to invest certain parts with the capability of acting upon the substance to be transformed or manufactured, in a certain definite manner, so that the result obtained can only be the desired one. Hence we see that the principal characteristic of a machine is its perfectly rigid and controllable manner of action.

3. Having defined the nature of the mechanical process and the machine, we have next to inquire into the nature of the information necessary to enable us to construct a machine for a specific purpose: this we shall find may be classified under four different heads, namely:—

a. A knowledge of the nature, both chemical and mechanical, of the raw material to be acted upon, as also of the product to be obtained by the process.

b. A knowledge of the nature of the motive power to be employed, and its most advantageous mode of action.

c. The influence of the laws governing bodies in motion, or dynamics, on the working of machinery.

d. A knowledge of the nature as to mode of action, requisite strength, and proportions of existing mechanical combinations.

4. To consider the first section generally would lead us too far for the limited nature of this book; it forms a complete science of itself termed "technology," and we must refer the reader to works on this subject for information. We shall, however, when describing machines for specific purposes, consider the nature of those raw materials with which they have to deal.

The other sections will form the subjects on which the following pages treat; and the author must beg to apologise for the incomplete nature of the same, on account of the limited space which this little book affords; trusting, however, that what it does contain may prove to the reader of some practical value.

CHAPTER I.

OF MOTIVE POWER.

5. WE understand by the term *motive power* a combination of the two elements *force* and *motion*. Every *force* or *pressure* is capable of producing motion; it implies a capability of continuing its effect, even when the object upon which it is exerted is moved by it from a state of rest. Where this motion does not take place, it is because the force is balanced by another acting with the same intensity in a contrary direction; or where it is exerted on an inert mass of matter, it is because the force is not of sufficient intensity to overcome the friction between the mass and the surface with which it is in contact.

But though every force may be capable of producing motion, yet we only recognise as motive power such forces as exist already in a state of motion; and to the sources of these especially, their nature and the methods adopted for their measurement as to intensity and quantity, we have now to direct our attention.

6. *Motion.*—The term *motion* implies a continual alteration of position in space; we have absolute and relative motion, as we have absolute and relative rest. Absolute motion is a change of position in space, relative motion a change of position relating to any other object, itself either in motion or at rest. The earth has an absolute motion; a man walking on board a vessel in motion has a relative

motion to the ship, and if he walk backward with the same speed as the vessel moves forward, his relative motion to the vessel is a state of rest relatively to the earth.

Motion, as to its nature, may be *uniform*, *accelerated*, or *retarded;* with respect to its direction, *rectilinear*, *circular*, *curvilinear*, *continuous*, *reciprocating*.

7. *Velocity* is motion considered in combination with time; the velocity of a body in motion is represented by the space it passes through, during a unit of time; thus, for uniform motion if s = space passed through, t = units of time, v = velocity, then:

$$1)\ v = \frac{s}{t} \qquad 2)\ s = v \times t \qquad 3)\ t = \frac{s}{v}$$

or: *the velocity is equal to the quotient of space passed through, divided by the corresponding time; the space passed through is equal to the velocity multiplied with the time*, and *the time is equal to the quotient of space divided by velocity.*

8. *Uniformly accelerated motion* is an equal increase of velocity for every unit of time. If in fig. 1, plate I., $a\ a'$ represent a velocity imparted to a body in a state of rest, during the infinitely small portion of a second represented by the line $a\ b$, then the space passed through will be represented by $a\ a' \times a\ b$, or the area of the rectangle $a\ a_1\ a_2\ b$; if further, at the end of this time the velocity be increased by the same amount $a_2\ b_1$, during the short time $b\ c$, then the total velocity during this time will be represented by $b\ b_1$, and the space passed through by $b\ b' \times b\ c$. Thus, continuing an increase of velocity during every particle of time, the body at the end of the first second, will have arrived at m with a final velocity g equal to $n \times a\ a_1$, if n represent the number of infinite particles of time in one second; and the space passed through will be represented by the sum of the infinitely narrow rectangles $a\ a^1\ a^2\ b$ &c., or the area of the triangle $a\ m\ n = \frac{a\ n \times g}{2}$, or (as $a\ n$ represents one

second) $= \frac{1 \times g}{2}$. At the end of the next second, the velocity of the body will be equal to 2 g, and the space passed through

$$= \frac{2 \times 2\,g}{2}$$

At the end of the third second the velocity will be equal to 3 g, and the space passed through

$$= \frac{3 \times 3\,g}{2}$$

or after t seconds the velocity

$$v = g\,t \quad . \quad . \quad . \quad . \quad . \quad \text{I.}$$

and the space passed through

$$S = \frac{t^2 \times g}{2} \quad . \quad . \quad . \quad . \quad \text{II.}$$

That is: *The velocity of a body, to which is imparted a uniformly accelerated motion by the continuous action of a certain force upon it, is equal to the product of the velocity which it attained at the end of the first second, with the number of seconds it has been in motion.*

The space passed through is equal to half the product of the square of the number of seconds, with the velocity attained at the end of the first second.

From equation I.: $t = \frac{v}{g}$ this value substituted in *eq.* II. gives:

$$S = \frac{v^2}{g^2} \times \frac{g}{2}$$

$$S = \frac{v^2}{2g} \quad . \quad . \quad . \quad . \quad \text{III.}$$

and from this:

$$v = \sqrt{2\,g\,s} \quad . \quad . \quad . \quad \text{IV.}$$

from *eq.* II.:

$$t = \sqrt{\frac{2\,s}{g}} \quad . \quad . \quad . \quad \text{V.}$$

The space passed through is equal to the square of the final velocity, divided by twice the increase of velocity per second.

The velocity at the end of a certain time is equal to the square root out of twice the product of the momentary acceleration, with the space passed through. The time which it has taken a body to move through a certain space is equal to the square root of the quotient of twice that space, divided by the momentary acceleration.

9. These equations express the laws of accelerated motion as caused by the continued action of a force of constant intensity upon a body in motion. Nature affords us an example of this description of motion in the action of the attraction of the earth on a falling body; for though, correctly speaking, the intensity of this force varies with the square of the distance from the centre of the earth, yet the spaces through which bodies fall are so small compared with this distance, that the force acting upon them may be considered as constant.

10. It has been ascertained by means of Atwood's apparatus for measuring the fall of bodies, that at the end of the first second the space passed through by a falling body is 16·08 feet, and therefore the velocity at the end of the first second = 32·16 feet for the latitude of London. That is to say, that if a body be continually acted upon by a force equal to its own weight, it will have a momentary acceleration of 32·16 feet per second. Now, as it is further known that the velocity imparted varies directly as the ratio of the mass of a body to the force acting on the same, we are enabled by this datum to find what the velocity and space passed through by a body will be, when acted upon by any other force than gravity.

For instance: A cannon ball weighing 6 lbs. is acted upon by the expansive force of gunpowder = 40,000 lbs. during a distance of 6 feet, the length of the bore, then g will be equal to

$$\frac{32 \cdot 16 \text{ ft.} \times 40,000 \text{ lbs.}}{6 \text{ lbs.}} = 214,400 \text{ ft.}$$

and from equation IV. the velocity with which the ball starts from the mouth of the gun is—

$$\begin{aligned} v &= \sqrt{2 \times 214,400 \times 6} \\ &= \sqrt{2,572,800} \\ &= 1604 \text{ ft. per second;} \end{aligned}$$

and the time which elapsed from the firing of the gun to the exit of the ball, from equation I.

$$t = \frac{1604}{214,400} = \frac{1}{134}\text{th of a second.}$$

A locomotive pulling horizontally at a train of carriages with a force = 8 tons, the weight of the train being 80 tons, will, at the end of the first second from the time of commencement of the motion, have imparted to the train a velocity—

$$g = \frac{8}{80} \times 32 \cdot 16 \text{ ft.} = 3 \cdot 2 \text{ ft.};$$

at the end of the first minute it will have gone a distance—

$$s = \frac{60^2 \times 3 \cdot 2}{2} = 5760 \text{ ft.},$$

supposing the locomotive to continue to pull with the force equal to 8 tons.

11. *Uniformly retarded motion* obeys the same laws, only in the reverse sense, as uniformly accelerated motion; that is, if a body is moving with a certain velocity, and there is a force continually acting upon it in a contrary direction, its motion will become uniformly retarded, and it will eventually come to rest. If v represent its original velocity, g the momentary retardation due to the opposing force, then at the end of t seconds, its velocity will be

$$v^1 = v - g\,t \quad . \quad . \quad . \quad . \qquad \text{I.}$$

and the space passed through

$$s = \frac{v^2 - v'^2}{2g} \quad . \quad . \quad . \quad . \quad . \quad \text{II.}$$

or

$$s = \frac{v^2}{2g} \quad . \quad . \quad . \quad . \quad . \quad . \quad \text{III.}$$

the distance passed through when it has come to rest.

Motion may also be irregularly accelerated or retarded, but the consideration of these descriptions of motion does not enter into our present object.

12. *Motive Power* is not tangible to our senses, except by means of its effect; and, therefore, our only method of estimating the same is by measuring the *amount of work* it is capable of performing; this *work performed* being in all cases reducible to the overcoming of a certain resistance or opposed force for a given distance in a certain time, and being equal to the power expended. Now, the only tangible idea we have of *force* is its relative intensity to that of the attraction of the earth: if a force balances the action of gravity upon a mass of matter whose weight is eight pounds it is said to be a *force* or *pressure* of eight pounds. Thus then, we estimate the intensity of motive power by reducing the work performed by the same, to a value expressed by a certain weight raised a certain height in a given time, and then measuring this quantity by means of a *unit of work*, consisting of the three elements, weight, distance, and time. This unit has been fixed for this country at the weight of one pound raised one foot high in a second, or, as it is commonly called, one *lb. ft.* For France, the unit is one *kilogramme meter*, or 2·2 lbs. raised 3·3 feet high in a second. A larger unit was introduced by James Watt for measuring the power of his steam-engine, which he called a *horse power*, and fixed at 550 lbs. feet per second, or 33,000 lbs. feet per minute. For instance: if we have a steam-engine, the cylinder of which is 2 feet in diameter, and

the stroke 3 feet, the piston moving at the rate of 50 strokes a minute, with an average pressure of steam equal to 10 lbs. on the square inch, we shall estimate the motive power of the same in the following manner:—

Area of piston, 2 feet diameter		= 452 sq. in.
Pressure on the same, per square inch . . .		10
Total pressure on piston		4520 lbs.
Length of stroke	= 3 feet	
Number of strokes per minute . .	50	
Distance the piston travels per minute . .		150 feet
Total amount of work performed per minute		678,000 lbs. feet.

678,000 : 33,000 = 20·6 H.P. of Engine.

13. This mode of measurement, though universally adopted, is, however, not theoretically correct for all cases; in fact, it is only so when motive power and resistance are in a state of *dynamical equilibrium*, hereafter to be explained, in which power and resistance are balanced *while in a state of motion*. Here we rightly express the amount of power employed by the distance in which the resistance is overcome; but when the resistance has to be moved from a state of rest, it is obvious that we can no longer express the work done by the pressure of the resistance multiplied with the distance passed through in a certain time; for if the pressure of the motive power is exactly equal to that of the resistance, and they are both in a state of rest, what is there that would call forth motion? Why, there must be an *excess of pressure* in the motive power, acting as an accelerating force to produce the motion required, and this *excess of pressure* would not appear in our method of calculation.

This not only applies to cases of moving from a state of rest; but, in general, to all instances where accelerated or retarded motion takes place.

14. To illustrate this, let us take, for example, Atwood's

apparatus (fig. 2, plate I.), consisting of two equal weights, say 3 lbs. each, suspended by a very thin thread over a pulley running on friction rollers. Considering the weight of the thread as nothing, these two weights will remain in a state of rest in whatever position we may place them. But if we place upon one of them a weight of 2 lbs., then the force of gravity due to the 2 lbs. will be acting upon a weight of 8 lbs., or four times the former amount, and will therefore, in one second, cause the two weights to move through a distance $= \frac{1}{4} \times 16$ feet $= 4$ feet. Now if, instead of placing on this additional weight, we give one of the weights an instantaneous impulse, so as to produce exactly the same effect, namely, to make the two weights pass through 4 feet in one second, the whole mechanical effect produced in the former case by the accelerating action of the 2 lbs. during the whole distance of 4 feet, will in this case be produced during the short duration of the impulse, and through the rest of the distance the weights will pass with an uniform velocity. Supposing the friction of the roller to be infinitely slight, the weights will in that case also pass through a distance of 4 feet during the lapse of another second; and we shall then have three cases, in each of which, according to the above means of measurement, we should express the mechanical effect or work performed by the raising of a weight of 3 lbs. through a distance of 4 feet in one second, or by $3 \times 4 = 12$ *lbs. feet*. But it is evident that this value is only correct as regards the last case, where the inertia of the descending weight causes it to continue to raise the other, but where the motion actually was produced by a previous separate effort. In the former cases, then, we have to add something to the apparent value of the resistance overcome (12 lbs. feet), to obtain a correct estimate of the power expended to produce that effect. This extra amount of power is expressed by a force which, acting continually upon the 3 lbs., will impart to them at the end of the first second a velocity equal to 8 feet.

Now if this force be represented by x, we shall have

$$\frac{32 \cdot 16 \text{ ft.} \times x \text{ lbs.}}{3 \text{ ft.}} = 8 \text{ ft.}$$

$$x = \tfrac{3}{4} \text{ lbs.}$$

and as these $\frac{3}{4}$ lbs. have to be exerted through a distance of 4 feet, they produce a mechanical effect equal to $\frac{3}{4}$ lbs. $\times$ 4 ft. = 3 lbs. feet, which amount, added to the 12 lbs. feet, will give 15 lbs. feet as the true amount of power expended in raising a weight of 3 lbs. through a distance of 4 feet, and imparting to it a final velocity equal to 8 feet per second.

15. It will be noticed that we have here no balance between *power expended* and *work performed*, as in our former example; the former being represented by 20 lbs. feet, and the latter only by 15 lbs. feet; but it is evident that, viewing it in this light, such must be the case, as we must always have a surplus of motive power to give the accelerated motion. Professor Robison, however, in his "Mechanical Philosophy," * indicates a method of obtaining an equation between the two by taking into consideration the amount of mechanical effect *extinguished* in each, and reasons thus: "The 5 lbs would, if gravity had acted freely upon it, have fallen 16 feet, but it has fallen only 4 feet, therefore a force has acted upon it sufficient to make it describe 12 feet in a second, with a uniformly accelerated motion; for it has counteracted so much of its weight. The thread was strained with a force equal to $3\frac{3}{4}$ lbs., or $\frac{3}{4}$ths of 5 lbs. In like manner the 3 lbs. weight would have fallen 16 feet; but it was raised 4 feet. Here was a change precisely equal to the other. A force of $3\frac{3}{4}$ lbs. acting on a mass whose weight is only 3 lbs., will in one second cause it to describe 20 feet with a uniformly accelerated motion. Now 5 $\times$ 12, and 3 $\times$ 20, give the same product, 60. And thus we see, that the quantity of motion extinguished or produced, and not the product of the weight and height, is

* Robison's "Mechanical Philosophy," art. "Machinery," pp. 226-27.

the true unequivocal measure of mechanical power really expended, or the mechanical effect really produced, and that these two are always equal and opposite."

16. As all machines have to be started from a state of rest, according to our present deductions the estimation of the power required to drive them would be considerably complicated if the work performed by them offered a *constant resistance*, or if the resistance were independent of the velocity of the machine. But fortunately this is only the case in a few instances, whereas in general the resistance offered at the working point increases in a certain proportion to the velocity of the machine: this enables the power, calculated for the full amount of resistance, to overbalance the same at the commencement of the motion, and gradually to accelerate the latter, till, the pressure of the motive force decreasing as the velocity increases, a period arrives when the pressure of the motive force and that of the resistance become equal, and a uniform motion ensues. There are however machines in which the resistance is acting to its whole extent at the beginning of the motion; such as for instance the steam hammer, where if the pressure of steam on the piston is only just equal to the weight of the hammer it will just balance the latter, but will not be able to lift it. In such cases we must therefore, as before stated, calculate our motive power for a larger resistance than would appear, were we only to take for its measure the weight multiplied with the height through which it is raised.

17. *The sources of motive power* may be divided into two classes:

(1.) *Primary sources*, or such where the motive principle is contained in the source itself.

(2.) *Secondary sources;* being those where the motive principle is imparted from some primary source, and which form a magazine in which the effect of the latter is accumulated.

To the first class we reckon:—

a. The attraction of gravitation.

b. The expansive force of gases.

c. Muscular force.

d. Magnetic attraction.

To the second class:—

a. The action of a body of water either as a falling body or as a stream.

b. The force of wind.

c. The pressure of the atmosphere.

18. The term *attraction of gravitation* is applied to that faculty which our earth possesses of attracting all masses of matter to its surface. This power of mutual attraction with which all matter is imbued, was found by Newton to vary directly as the masses, and inversely as the square of the distance; or, if a mass equal to 1 at a distance equal to 1 is attracted with a force equal to f by another mass also equal to 1, then it will be attracted by a mass = M at a distance 1 with a force $= f \times M$, and at a distance $= d$ with a force $= \frac{f \times M}{d^2}$.

A mass m will be attracted by a mass M at a distance d with a force

$$F = \frac{f\,M\,m}{d^2}.$$

For instance: the mass of the sun is 355,000 times greater than that of the earth, therefore, a body on its surface would have a weight 355,000 times greater; but again, the diameter is 112 times larger, and therefore the weight would be decreased $112 \times 112 = 12{,}544$ times. The actual increase of weight would therefore only be $\frac{355{,}000}{12{,}544}$ or about $28\frac{1}{3}$ times.

19. The *mass* of a body is generally considered to be synonymous with its weight; this is clearly an error, as the *mass* is always a constant quantity wherever it may be, but the

weight of it will vary according to the laws just explained. As we do not possess any other means of judging of the mass of a body than by its weight, and as the accelerating force g obviously varies with the variation of weight in one and the same mass, the amount of the latter is generally expressed by $M = \frac{weight}{2\,g}$ which will always give a constant quantity.

We have before stated that the general measure for the mechanical effect contained in a body in motion, is the product of the force acting on that body with the distance passed through. Now, according to equation III. § 8, a body moving with a velocity $= v$ is capable of overcoming a resistance, which would diminish its velocity by g feet every second, through a distance equal to $S = \frac{v^2}{2\,g}$, and as for a weight $= 1$, a resistance $= 1$ will diminish its velocity by 32·16 feet per second, then the mechanical effect which the weight of one pound is capable of overcoming when moving with a velocity $= v$ feet per second, is equal to $\frac{v^2}{2 \times 32{\cdot}16} \times 1$ lb. = lbs. feet per second; or the mechanical effect contained in a weight P will be

$$\text{I.} \quad . \quad . \quad V = \frac{v^2 \times P}{2 \times 32{\cdot}16} \text{ lbs. feet per second;}$$

and as we represent the mass of a body by $M = \frac{P}{2 \times g}$ then

$$\text{II.} \quad . \quad . \quad V = v^2 \times M.$$

This is generally termed the *vis viva* of a body in motion, and this, it will be observed, is identical with the value expressed by 3 lbs. feet in § 14, being the effect or *vis viva* of the weight of 3 lbs. moving with a velocity of 8 feet per second, in a contrary sense to the action of gravity.

We will illustrate by one or two examples the method of

calculating the *vis viva* of bodies in motion, or the amount of accumulated work contained in them.

20. The force which it is necessary for a locomotive to exert in pulling horizontally at a train of carriages to enable it to overcome all the resistances due to friction, &c. is nearly proportional to the total weight of the train. If g represent this weight, then $\frac{g}{n}$ will represent the force required where n is a constant, whose value is generally $= 200$. As long as there is only exerted, in the cylinders, a pressure P equal to the resistance $\frac{g}{n}$, no motion will ensue, but if this pressure become $P + p$, the train will be started, and, according to the value of the increment p, will move with an increasing velocity, which, after $P + p$ has acted during a certain time t, shall have become v. This velocity is the same as would have been originated in a mass $\frac{G}{2\,g}$ upon which a constant force of p only had acted, and the *vis viva* accumulated in the train due to this velocity will be $\frac{G}{2\,g}v^2$, which represents the total effect of the pressure p during the time t. During this time a distance S will have been passed through; therefore the resistance $\frac{G}{n}$ has been overcome through a distance S, representing a mechanical effect of $\frac{G}{n}$ S, produced by the pressure P during the time t. Therefore, the total mechanical effect produced by the total pressure of the steam during the time t is represented by the sum of the two separate effects:

$$\frac{G}{m}S + \frac{G}{2g}v^2 \quad . \quad . \quad . \quad \text{I.}$$

At the end of the time t, from some cause or other, the surplus pressure p ceases to be exerted, and P being equal to the resistance $\frac{G}{n}$ of the train, the same will continue henceforth to move with a velocity $= V$. If thus during the time t', it passes through a distance S', the mechanical effect of the pressure P during this time will be represented by $\frac{G}{n} S'$.

At the end of the time t' the pressure of steam is cut off altogether. The resistance $\frac{G}{n}$ will, therefore, now retard the speed of the train until it is brought to a stand-still, which will take place at the end of a certain time t'', during which time a distance S'' will have been passed through; and the mechanical effect $\frac{G}{n} S''$ thereby produced, will be clearly equal to the *vis viva* $\frac{G}{2g} V^2$ contained in the train at the commencement of the time t''. We have therefore:—

$$\frac{G}{n} S'' = \frac{G}{2g} V^2 . \quad . \quad \quad . \text{ II.}$$

from which equation we are enabled to find the distance S'' that the train will go after the steam is cut off: $S'' = n \frac{V^2}{2g}$, that is: n times the height corresponding to the velocity V. Thus, if $V = 50$ feet and $n = 200$, then $S'' = 200 \frac{50^2}{2 \times 32} =$ 7812 feet.

21. As a second example we will consider the action of a cannon ball when fired from a gun. Here we must neglect: *a*, the friction of the air; *b*, the friction of the ball in the bore of the gun; *c*, the vibrations that take place in gun and ball; and we assume: *a*, that the pressure of the gunpowder when ignited is exerted equally in every direction; *b*, that there is nothing to prevent the gun-carriage recoiling. When the gun is fired, the expansive pressure of the

generated gas is exerted with the same intensity on the back of the chamber of the gun as it is on the ball, and, were they both of equal mass, both would be started with the same velocity in opposite directions. As it is, however, the mass of the gun and carriage being many times greater than that of the ball, the velocities of ball and gun will be inversely proportioned as their respective masses, that is, if M the mass and V the velocity of the ball, and $M_{,}$ the mass and $V_{,}$ the velocity of the carriage and gun, then,

$$M V = M_{,} V_{,} \quad . \quad . \quad . \quad . \quad . \quad \text{I.}$$

The proportion of M to M′ being generally as 1 to 300, the velocity with which the ball starts will be about 300 times that of the carriage.

The mechanical effect produced in the ball will be $M V^2$, that produced in the carriage $M_{,} V_{,}^2$, and their proportion will be $\frac{M V^2}{M_{,} V_{,}^2}$, or, if we substitute the value of $V_{,} = \frac{M}{M_{,}} V$ out of equation I.:—

$$\frac{M V^2}{M_{,} V_{,}^2} = \frac{M V^2}{M_{,} \frac{M^2}{M_{,}^2} V^2} = \frac{M_{,}}{M}$$

that is: *the vis viva contained in the ball is to that contained in the carriage, as the mass of the carriage is to that of the ball.*

For a 24-pounder gun the weight of the ball is 24 lbs. that of the gun and carriage about 7200 lbs., and the velocity with which the ball leaves the gun about 1600 feet.

Therefore:

$$M = \frac{24}{64}, V = 1600, M_{,} \frac{7200}{64}$$

The mechanical effect produced by the ball will be

$$\frac{24}{64} \times 1600^2 = 960,000 \text{ lbs. ft.} = 1745 \text{ H.P.}$$

that produced by the carriage:

$$M V^2 \times \frac{M}{M_{\prime}} = \frac{1745}{300} = 5 \cdot 8 \text{ H.P.};$$

and the velocity with which the latter runs back:

$$V_{\prime} = V \frac{M}{M_{\prime}} = \frac{1600}{300} = 5 \cdot 3 \text{ feet.}$$

22. *The expansive force of steam* or of gases in general, constitutes our most concentrated form of motive power. This expansive property is caused by the repulsive action of the molecules when under the influence of heat. Mariotte discovered, that with most gases this force varied inversely as their density, supposing the temperature to remain the same. That is: If a volume of steam exerts a pressure of 20 lbs. on the square inch, and it is permitted to expand into twice that volume, its pressure will be reduced to 10 lbs. on the square inch. This law is, however, somewhat modified by the fact that the heat contained in the gases varies with their pressure.

Steam possesses one great advantage over atmospheric air and other gases used as motive agents; namely, its perfect condensability, that is, its capability of reduction to the form of water by the extraction of the heat contained in it, thus affording, over and above the force contained in itself, an extra pressure equal to that of the atmosphere, or nearly 15 lbs. on the square inch. The perfect attainment of this condensation, together with the working of high pressure steam down to a very great degree of expansion, afford the maximum of economy attainable with this powerful motive agent. But this employment of steam expansively has one great disadvantage; namely, the inequality of the pressure on the piston throughout the stroke, owing to its gradual decrease as the steam expands in the cylinder. To overcome this difficulty three methods have been adopted:—

a. The employment of the steam in such a manner that

it acts first at a high pressure and expands slightly in a small cylinder, and then passes into a larger cylinder, where it is allowed to expand to a low degree. Thus, by dividing the whole expansion over two strokes, the variation of pressure at the beginning and end of each stroke is considerably diminished, and by having two cylinders, one high- and the other low-pressure, acting together, the power of the up and down stroke is equalised, which would not be the case had we only one cylinder in which the steam first acted at a high pressure on one side of the piston and then at a low pressure on the other side.

b. By expending the surplus power resulting from the high pressure of the steam at the commencement of the stroke, in increasing the velocity of a heavy fly-wheel; power is thus accumulated in the latter, which is given off towards the end of the stroke, when the pressure on the piston has considerably diminished. We shall consider this method further when speaking of the action of the fly-wheel in particular.

c. By coupling a pair of engines at right angles; the steam will then act with high pressure at the commencement of the stroke in one cylinder, while the other cylinder is at half stroke, and consequently the steam in it at a low pressure. This method would become the more efficacious, the greater the number of engines there were coupled together in such a manner, that each commenced action at a fractional part of the stroke.

23. The following Table indicates the corresponding temperature to different pressures of steam, and also its volume as compared with that of the water which generated it :—

Pressure in lbs. per square in.	Corresponding Temperature, Fahr.	Volume of steam, that of the water being = 1.	Pressure in lbs. per square in.	Corresponding Temperature, Fahr.	Volume of steam, that of the water being = 1.
1	102·9	20954	51	283·6	542
2	126·1	10907	52	284·8	532
3	141·0	7455	53	286·0	523
4	152·3	5695	54	287·2	514
5	161·4	4624	55	288·4	506
6	169·0	3901	56	289·6	498
7	176·0	3380	57	290·7	490
8	182·0	2985	58	291·9	482
9	187·4	2676	59	293·0	474
10	192·4	2427	60	294·0	467
11	197·0	2222	61	295·0	460
12	201·3	2050	62	296·0	453
13	205·3	1903	63	297·1	447
14	209·0	1777	64	298·1	440
15	213·0	1669	65	299·1	434
16	216·4	1572	66	300·1	428
17	219·6	1487	67	301·2	422
18	222·6	1410	68	302·2	417
19	225·6	1342	69	303·2	411
20	228·3	1280	70	304·2	406
21	231·0	1224	71	305·1	401
22	233·6	1172	72	306·1	396
23	236·1	1125	73	307·1	391
24	238·4	1082	74	308·0	386
25	240·7	1042	75	308·9	381
26	243·0	1005	76	309·8	377
27	245·1	971	77	310·7	372
28	247·2	939	78	311·6	368
29	249·2	909	79	312·5	364
30	251·2	882	80	313·4	359
31	253·1	855	81	314·3	355
32	255·0	831	82	315·2	351
33	256·8	808	83	316·1	348
34	258·6	786	84	317·0	344
35	260·3	765	85	317·9	340
36	262·0	746	86	318·7	337
37	263·7	727	87	319·5	333
38	265·3	710	88	320·3	330
39	266·9	693	89	321·1	326
40	268·4	677	90	321·9	323
41	269·9	662	91	322·7	320
42	271·4	647	92	323·5	317
43	272·9	634	93	324·3	313
44	274·3	620	94	325·1	310
45	275·7	608	95	325·9	307
46	277·1	596	96	326·6	305
47	278·4	584	97	327·3	302
48	279·7	573	98	328·0	299
49	281·0	562	99	328·7	296
50	282·3	552	100	329·4	293

The pressure indicated in the Table is exclusively that of the atmosphere, that is to say, it represents the actual pressure in the boiler as indicated by the steam gauge.

From this table we see the great economical advantages as regards the consumption of fuel, of employing steam at a high pressure; for whereas to raise the pressure of steam at 7lbs. by 1lb we require an increase of temperature of 6 degrees Fahrenheit, we only require to raise the temperature of steam of 70lbs. pressure by 1 degree, to obtain the same increase of pressure.

24. If we represent in fig. 3, plate I., the different pressures of steam by ordinates, and the corresponding volumes occupied by it at these different pressures, by horizontal divisions, we obtain a curve which is termed the expansion-curve of steam, and by means of the area of the figure produced (generally called the "indicator figure of the expansion of steam") we are enabled to calculate the exact duty which steam, allowed to expand at a certain pressure to a given degree, is capable of performing.

If each of the horizontal divisions represent 100 volumes of steam, then, according to the figure, we find that steam at 2lbs. pressure occupies 109 divisions, or 10,900 volumes, which corresponds closely enough for practical purposes with the figures in the table, Again, steam of 71 lbs. occupies 4 divisions, or 400 volumes as shown in the table.

To arrive at the duty performed by steam of a certain pressure, when expanded to a certain degree, by means of this diagram, we proceed in the following manner:—

Supposing we have steam of 82lbs. pressure, and wish it to expand into five times its original bulk, that is, we cut off the communication between cylinder and boiler when the piston has performed $\frac{1}{5}$th of the stroke, then by the diagram we find that steam at 82lbs. pressure occupies $3\frac{1}{2}$ divisions. These multiplied by 5 will give $17\frac{1}{2}$ divisions, corresponding to a pressure of 14lbs. which will be exerted by the steam when expanded; and if we allow a vacuum equal to 12lbs. pressure obtained by condensation, the area of the figure enclosed between the horizontal lines at 3 and 82lbs. and between the vertical divisions ending at 17·5, will represent the duty performed by steam at 82lbs. pressure expanded 5 times and condensed. The average pressure will be obtained by dividing the number of divisions contained in this figure by the number of horizontal divisions at the 14lbs. The number of divisions contained in the area between every lb. pressure as computed from the table, are noted on the left hand side of the diagram; in counting these for our indicator figure, we find they number 676, which divided by 17·5, will give 39lbs. as the average pressure throughout the stroke. By multiplying this by the area of the piston and by the length and number of strokes per minute, and dividing the result by 33,000, we obtain the horse-power of the engine.

25. By means of an instrument termed the indicator, a diagram similar to this would be traced by the pressure of the steam itself, had we a steam-engine constructed to work with the degree of pressure and expansion here determined on. Fig. 4, Plate I., represents such an "indicator-figure," as it is termed, actually taken from the Crystal Palace pumping engines, hereafter described. It shows, not only the duty performed by the engine, but also in what manner it is performed; that is, whether the engine is in proper working order. To show how this is, we must first give a short description of the indicator itself, and of its mode of action.

Fig. 5, Plate I., represents the indicator; *a* is a cylinder, in the lower part of which is a closely fitting piston, made to work in the cylinder without much friction. The underside of this piston is in direct contact with, and is acted upon by the steam in the cylinder to which the indicator is attached. The upper side is in connection with a spiral spring, fixed in the upper part of the cylinder, in such a manner that the piston, in moving up beyond the division on the scale *g*, marked *o*, has to compress it; but in moving down below that mark, has to extend or pull at the spring; therefore, when the piston is at the mark *o*, the spring is *in equilibrio*. As the upper side of the piston is in direct communication with the pressure of the atmosphere by means of the slot *b*, it is evident that the piston can only be moved below the mark *o* by this pressure, or, in other words, by a vacuum below the piston, while all pressure indicated above *o* will represent that of the steam. Through the slot *b* the piston is further attached to the index and pencil *c*, which consequently moves up and down with the same. Round the cylinder *a* is fixed a large cylinder *d*, in such a manner that it can only move round *a* to a certain extent, being brought back to its original position by a spring. The motion backward and forward of this cylinder is effected by the string *e*, one end of which is attached to it and the other end to some part of the engine that moves in a similar manner to the piston. A piece of paper is fixed round this cylinder by means of the clip *f*, upon which the pencil *i* traces, as it moves up and down, a line indicative of the pressure in the steam cylinder at consecutive parts of the stroke. The tap *h* opens or cuts off the communication between the steam in the cylinder and the piston of the indicator.

In referring to Fig. 4, we find a line drawn termed "atmospheric line;" this is drawn by the pencil of the indicator, when the spring is *in equilibrio* and the index at *o* on the scale, by moving the string attached to the outer cylinder by hand. This line divides the diagram into two parts: the upper shows the actual pressure of the steam; the lower, the pressure obtained by forming a vacuum in the condensor. The figures on the right hand side give these two pressures separately, while those on the left hand side indicate the total pressure obtained. The manner in which this diagram is formed is the following:—

At the commencement of the stroke the pencil is situated

at the bottom of the figure at the left hand, and the moment steam enters the cylinder, the pencil is raised suddenly to the top of the figure, describing the vertical line 2—32. As long as the steam is full on, it will continue at this height and describe the horizontal line 0—3, in consequence of the paper moving under the pencil with the stroke. When the steam is cut off and begins to expand, the pencil will gradually drop as the pressure decreases, describing the curve 3—10, which corresponds with that of fig. 3. Arrived at the end of the stroke, the exhaust port is opened, and the pencil drops suddenly, and if the vacuum formed by condensing the steam were complete, it would fall to 14·7; but as this is practically almost impossible, it falls till the difference of pressure of the atmosphere and of the spring amounts to whatever pressure there is left in the condensor, *i. e.*, about 3 lbs. During the up stroke the outer cylinder moves back to its original position, the pencil tracing the line 10—0 on the paper. We perceive that this line drops slightly towards the end of the up stroke, which shows that the vacuum becomes more complete during that time; in fact, we find that the 3 lbs. back pressure at the commencement has been reduced to $1\frac{1}{2}$ lb. Arrived at 0, the steam is admitted on top of the piston, and the pencil moves as before.

26. In order to show in a more striking manner how by means of the indicator-figure we may accurately ascertain the manner in which the engine is working, we will contrast Fig. 6, Plate I.—the indicator-figure taken off an engine in which the slide-valve motion had become damaged—with the diagram Fig. 4. The line 2—32 in this figure is represented by the line *a b c* in fig. 6. The part *a b* commencing to rise before the end of the stroke, indicates that the exhaust port is closed too soon, while the part *b c* shows, by attaining its greatest height only after one tenth of the stroke is passed through, that the steam is admitted too slowly, or, as it is

termed, the steam is "wiredrawn." The line *c d* indicates that the steam is only allowed to expand after two-thirds of the stroke have been passed through, *e f* shows that the exhaust port is opened too soon, although, as indicated by *f a*, this opening takes place so slowly that the steam is again wiredrawn on its passage to the condensor. It would lead us too far to consider this interesting subject further, we must therefore refer the reader to larger works on the steam engine * for more information, as also to a work by Mr. Hopkinson, containing practical information on the use of the indicator.

27. In referring to the table of pressure and expansion of steam § 23, we find that the greater the pressure the smaller is the volume as compared with that of the water from which it is generated; or in other words, in the formation of steam in the ordinary manner, a certain invariable amount of water must always be evaporated to obtain steam of a certain pressure. If however we heat this steam, which we may term *saturated* to a greater degree, when it is no longer in connection with the water in the boiler, then we shall have raised its pressure without having had to evaporate the extra amount of water due to that pressure, thus obviously effecting a saving in the amount of fuel consumed. Steam thus formed is termed *superheated*, and may be employed either in the above manner, that is, the *saturated* steam disconnected with the water and steam in the boiler and *superheated* (a greater pressure than exists in the boiler being thus obtained), or else the steam may be *superheated* on its passage to the cylinder while in connection with the steam in the boiler, by which means the pressure will remain the same, the steam expanding to a greater volume, thus giving a greater quantity of steam of the same pressure generated from the same quantity of water.

Experiments on the latter principle have recently been made, the results of which show a saving effected in the

* Tredgold, Clark's "Railway Engineering," &c.

fuel of from 20 to 30 per cent. as compared with that necessary for the formation of saturated steam of the same pressure.

28. *Muscular force* is the most ancient motive agent; its employment is now generally limited to domestic purposes and machines where but a small amount of power is required. Both the human and animal muscular force vary in their capability, according to the manner in which they are employed and the duration of the employment. Thus, for instance, Morin found that man, working eight hours a day, will perform the following units of work per minute :—

Raising his own body	4250 lbs. ft.
Working at the treadmill	3900 " "
Drawing or pushing horizontally . . .	3120 " "
Turning a handle.	2600 " "

Working six hours a day :—

Raising materials with a pulley	1560 " "
Raising materials with the hands	1470 " "
Raising materials upon the back and returning empty	1126 " "

Working ten hours a day :—

Raising materials with a wheelbarrow on ramps .	720 " "
Throwing earth to the height of five feet . .	470 " "

The force exerted by a horse at various rates of speed is as follows :—

Rate, in miles per hour .	2	3	$3\frac{1}{2}$	4	$4\frac{1}{2}$	5
Force exerted, in lbs.. .	166	125	104	83	$62\frac{1}{2}$	$41\frac{2}{3}$

According to this, his greatest mechanical effect will be produced when moving at a rate of 3 miles per hour, viz. 33,000 lbs. feet per minute; but this effect, according to Tredgold, he can only produce when working six hours a day.

Tredgold gives the following formula for ascertaining the pull of a horse when going at a certain speed :—

$$P = \frac{V-v}{V} \times 250,$$

where P = the pull of the horse in lbs., V his maximum velocity when unloaded, and v the velocity when loaded, at the given duration of labour per day, both in miles per hour.

Euler gives :—

$$P = \left(1 - \frac{v}{V}\right) K,$$

where K represents the greatest force a horse can exert at a dead pull, and the other values as above.

29. *Magnetic attraction,* as produced by electric or galvanic currents in what are termed *electro-magnets,* has of late years been much experimented upon with a view to ascertain to what extent the same might be employed as a motive agent; but owing to the great cost, as compared with steam power, of the materials consumed in its production, it has up to the present time not proved of any practical value. Until the maximum power of the magnet is gained, the same increases with the square of the strength of the electric current. But this power of attraction decreases very rapidly with the distance of the object; it has been found that at a distance of $\frac{1}{182}$nd of an inch, it was only one-third of the total attractive force when in contact.

30. *Water when actuated by gravity* forms, in mountainous districts, the most economical source of motive power, and stands next in importance to steam power. Its mode of action is twofold: either as a stream, where its force is proportional with the square of its velocity; or as a falling column, where its effect is measured by the product of its weight with the height of the column.

In a stream, the velocity which the water obtains is due to the inclination of its bed, or its *rate of fall.* This velocity would, were it not for the friction of the water on the sides and bottom, be equal to the amount due to a falling body, which had descended through a height equal to the total fall of the stream; but the friction reduces this velocity considerably, and the effective power of the stream is therefore not calculable from the total amount of its fall, but only from that amount due to its actual velocity; or if $V =$ this velocity, $W =$ the weight of the moving mass of water per second, and $g =$ the momentary acceleration, we have the total effective power expressed by—

$$\text{I.} \quad . \quad . \quad P = \frac{W \times V^2}{2g}$$

or, if the quantity of water is expressed by Q cubic feet per second:—

$$\text{II.} \quad . \quad . \quad P = \frac{62{\cdot}3\,Q \times V^2}{2g}$$

For instance: a stream delivers 200 cubic feet of water per second, moving at the rate of 8 feet per second; the total power contained in the same will therefore be—

$$P = \frac{62{\cdot}3 \times 200 \times 8^2}{2 \times 32} = 12{,}460 \text{ lbs. feet per second,}$$

which, divided by 550 lbs. feet, is $= 22{\cdot}6$ H. P. If, instead of allowing the stream to follow its natural course, we cause it to run in an artificial one which is constructed horizontally from a point at a high level of the stream, we shall at any point of this aqueduct have a vertical distance, equal to the whole amount of fall of the natural bed of the stream, due to the length of the artificial one, through which the water can fall by the unimpeded action of gravity; and the total power contained in the stream, if employed in this manner, will be $P = 62{\cdot}3\ Q \times h$, where Q is the quantity of water per second, and h the vertical height through which it falls. If we take the same stream as above, and consider the velocity of 8 feet per second due to a rate of fall of

20 feet per mile, and construct an aqueduct from a point in the stream 3 miles distant from where the power has to be applied, then the total available fall at this point will be $3 \times 20 = 60$ feet, and consequently the power obtained from the stream will in this case be—

$$P = 12,460 \text{ lbs.} \times 60 \text{ feet} = 747,600 \text{ lbs. feet,}$$

divided by $550 = 1356$ H. P., in the place of only 22·6 H. P., if employed as before.

31. Owing to its liquid yet almost incompressible nature, water has of late years been made the vehicle for the most concentrated form of motive power yet known, in its application to hydraulic presses, hydraulic cranes, &c. The principle upon which its great efficacy in these cases depends, is that of the perfectly equal distribution of a pressure throughout its whole mass, which is exerted upon only a very small portion of a body of water in a confined space; that is, if we have a cylinder of 1 square foot sectional area, and connected with the same a vertical pipe of 1 square-inch sectional area and 100 feet high, filled with water, we should have exerted upon each square-inch area of the piston in the cylinder, a force of 6232 lbs., and upon the whole piston therefore the pressure of $144 \times 6232 = 897,408$ lbs. $= 400$ tons. The same result would be obtained if, instead of making the pipe 100 feet high, we were to make it only long enough to admit a small piston loaded to the extent of 6232 lbs. We shall consider this subject more in detail when speaking of the machinery specially adapted to this description of power.

32. The *wind* is employed as motive power, particularly in flat countries, where there is a scarcity of water power, and where the employment of steam power would be too expensive.

Gravity may be also here considered the source of power, the cause of the motion of the air being traceable to its being forced by pressure from above to displace a volume which has been rarified by heat, and is thus caused to ascend. The pressure of the wind is considered to be proportional to the square of its velocity. If F = the force of the wind in pounds, $v =$ the velocity in feet per second, $s =$ the surface against which it presses in square feet, then—

$$F=\frac{v^2 s}{440}.$$

And if the surface upon which the wind strikes is inclined to the line of its direction by an angle $= \beta$, as is the case in windmills, then

$$F = \frac{v^2 s \sin.^2 \beta}{440}.$$

The following table gives the force of the wind when moving at different velocities upon the surface of one square foot:—

V		$F=\frac{v^2 s}{440}$	V		$F=\frac{v^2 s}{440}$
Miles per hour.	Feet per second.		Miles per hour.	Feet per second.	
		lbs.			lbs.
1	1·47	·005	30	44·01	4·429
2	2·93	·020	35	51·34	6·027
3	4·40	·044	40	58·68	7·873
4	5·87	·079	45	60·01	9·963
5	7·33	·123	50	73·35	12·300
10	14·67	·492	60	88·02	17·715
15	22·00	1·107	80	117·36	31·490
20	29·34	1·968	100	146·70	49·200
25	36·67	3·075			

33. The *pressure of the atmosphere* is employed as an auxiliary power in those steam-engines where a vacuum is formed on one side of the piston by condensing the steam. Its force amounts to about 15 lbs. on the square inch, and is equal to the pressure of a column of water 33 feet high, or a column of mercury 30 inches in height. In the first so-called steam-engines, such as Newcomen's, Savery's, &c., this atmospheric pressure was the sole moving agent, inasmuch as the steam was only employed to form a vacuum

under the piston, which was then caused to descend by the pressure of the air upon it.

34. The force and rapidity with which air rushes into a vacuum was made use of some years ago, in an endeavour to apply it as the motive power for railway locomotion; where the carriages were attached to a piston, impelled along, through a continuous cylinder, or pipe, fixed to the permanent way, on one side of which piston a vacuum was created, the other side receiving the pressure of the atmosphere. Owing to the impracticability of forming a perfectly air-tight closing of the slit along the top of the pipe, through which the connection between the piston and the carriages had to pass, the scheme was, however, found an unsuccessful one, and was abandoned. A modified form of the same has, however, been found to work with success in the United States, as applied to the transport of mail-bags to and from the principal post-office; the piston itself forming here the means of transport, into which the mailbags are deposited, and thus conveyed under ground through the streets to the different stations.

CHAPTER II.

OF THE MANNER IN WHICH THE MOTION OF MACHINERY IS INFLUENCED AND MODIFIED BY DYNAMICAL LAWS.

35. The motion of a machine, as regards its velocity, may be of a threefold nature; it is either *accelerated, uniform*, or *retarded;* and these three phases will occur under certain conditions with almost all machines. The manner in which each is brought about is dependent:

1. On the mode of action of the impelling power.
2. On the mode of the resistance at the working point.

The necessity for the occurrence of each of these phases of motion in a machine at one period or another will be seen by the following example:—

Taking the case of a flour-mill, driven by water-power, we will allow the water to flow on to the water-wheel for a certain time, and then, shutting off the supply, examine what will be the nature of the motion of the mill at different periods during this time.

When the water first flows on to the wheel, no motion in the same can ensue, as we have shown elsewhere, until the pressure on the wheel is greater than the resistance at the working point, *plus* the inertia expressed by the friction of the machine. As soon as this takes place, the mill will begin to move. Its motion being at first slow, and the actual resistance at the working point being consequently very small, compared to what it would be when working at full speed, the power at first expended in moving the mill will be almost entirely employed in overcoming the friction of the various moving parts.

As we have necessarily a surplus of power over and above the resistance of the mill, in order to bring about motion in the same, this surplus, when once the mill is in motion, will be employed in increasing its velocity. The motion of the mill will therefore, from the commencement, be *accelerated.*

36. We have before observed, that the pressure exerted by motive power is proportionate to the speed with which it has to follow up the moving resistance. If, therefore, the speed of the mill is increased, and consequently also that of the circumference of the water-wheel, the pressure of water upon the same, and, therefore, also the surplus of power, will decrease in like manner. This decrease of pressure will continue until the mill shall have attained its highest speed; the surplus of power being now completely absorbed, the resistance of the mill will exactly equal the pressure of the water exerted on the wheel. There being no further cause for an alteration in the state of things, the mill will henceforward move with a *uniform velocity*, and as there is now a balance between power and resistance, precisely similar to that which takes place just before the mill begins to move, we term this phase of motion the *dynamical equilibrium* of the machine.

If after a certain time, we cut off the supply of water to the wheel, the pressure of the motive power suddenly ceases, and, were the mill devoid of mass, it would as suddenly come to a stand still. But we know that a certain amount of power was at the commencement of motion spent in accelerating the moving parts, this power is still contained in them in the form of *vis viva,* or as it is termed in this case, *momentum of inertia,* and the mill will continue in motion until the whole of this is given out again. As this amount of accumulated power becomes gradually less, so will also the speed of the mill; therefore, from the time when the supply of power was cut off, the mill will move with *retarded velocity*.

37. In Fig. 7, Plate I., we have delineated graphically both the velocity of the mill and the intensity of the moving power during the whole period of action. The periods of time form the *abscissa*, and the velocity and pressure are drawn as ordinates; thus we get a curve *a b c d* representing the action of the motive power, and *a′ b′ c′ d′* representing the corresponding velocities in the mill. The portion *a′ b′* represents the duration of the acceleration, *b′ c′* that of the equilibrium, and *c′ d′* that of the retardation. The curve *a b* represents the intensity of the surplus of power during the acceleration; it begins from nothing at the commencement of motion, increases rapidly during part of the acceleration, when the motion is yet slow, comes to a climax when the quantity of water that can flow during a certain time on to the wheel reaches its maximum value, and then decreases as the velocity of the mill increases; from *b* to *c* it remains constant, and at *c*, when the supply is cut off, it drops suddenly to nothing.

38. From what has been said above it is clear then, that a period of dynamical equilibrium must occur at some time or other during the motion of every machine. The length of its duration, and the regularity of its occurrence, will vary in different machines according as the nature of their resistance and that of the motive power that drives them, varies. From the manner in which it occurs in different machines, we may divide these into three classes:—

1. Machines with constant equilibrium.
2. Machines with regular periodical equilibrium.
3. Machines with irregular periodical equilibrium.

To the first class belong principally such machines as have only rotatory motion in their moving parts, and in which the resistance at the working point remains constant; as for instance, the flour-mill. To the second class may be reckoned, those machines in which the resistance is, at regular periods, a constant quantity, or which have a "working stroke" and "returning stroke," as Dr. Robison terms them. As type of this class, we may consider the saw-mill, in which the time of the down stroke of the saw while cutting is the period of equilibrium.

To the last class belong all machines the resistance at whose working points are constant for irregular periods of duration, and are either constant or variable in intensity at these different periods. As type of this class we may consider the rolling-mill, in which the resistance is constant during the time that the iron is running through the rollers, but varies both as to duration and intensity, according to the number of times the iron has passed through and increases in length.

39. In each of these three classes, the action of the machine will be more or less modified by an alteration in the amount or description of any of the following elements :—

1. The amount of motive power acting upon the receiver.
2. The amount of resistance at the working point.
3. The amount of mass contained in the moving parts.
4. The geometrical construction of the various moving parts.

We will now briefly consider in what manner an alteration in each of these elements will influence the motion of each class of machine.

To represent *the first* class we will again take the flour-mill, to which we have before referred and consider,

40. (*a*). *The influence of the increase of motive power.* Let us suppose that in the first instance a quantity of 10 cubic feet of water per second was allowed to flow on to the water-wheel, and that during the equilibrium this quantity of water had to exert a total pressure of 2 cwts. on the circumference of the wheel. We now allow double that quantity of water, that is, 20 cubic feet per second, to flow on to the

water-wheel, leaving the resistance at the grindstones the same as it was before. This latter being the case, it is clear that motion in the mill will commence, as before, when the pressure at the circumference of the water-wheel just exceeds 2 cwts., which will be the pressure exerted during the equilibrium. As we have double the quantity of water, however, the wheel must now move twice as fast as it did in the first instance during the *equilibrium*, for this pressure to be exerted. The duration of the *accelerated motion* would be decreased to about one-half, were the wheel to move with the same velocity as in the first instance; but it would be increased to about double, were the amount of power the same as in the first instance, but the speed double. It will therefore remain about the same as it was, in consequence of the increase in both speed and power. As the mill goes twice as fast as it did in the first instance during the equilibrium, it will consequently grind twice as much corn. In consequence of this increase of speed the momentum in, or the amount of power absorbed by, the moving parts will be considerably greater, and therefore the duration of the retarded motion, commencing when the supply of water is cut off, will also be considerably greater than in the first instance.

41. (*b*). *The influence of the amount of resistance at the working point.* We now let the quantity of water flow into the wheel as in the first instance, namely, 10 cubic feet, but decrease the work performed by the grindstones by one-half, so that motion will now commence when the pressure of the water upon the wheel just exceeds 1 cwt. The results will be analogous to those arrived at in the second instance; that is, the velocity during the accelerated motion will be greatly increased as compared with the first instance, as also the velocity during the equilibrium, because the same quantity of water has to exert only half the pressure on the wheel, to overcome the resistance.

42. (*c*). *The influence of the amount of mass in the moving*

parts. Leaving everything else as it was in the first instance, we introduce upon one of the shafts of the mill a heavy fly-wheel. This, when once moving at a uniform speed, will not increase the resistance of the mill beyond a small amount due to extra friction. The pressure of water on the water-wheel will therefore remain 2 cwts. during the equilibrium; and as the same quantity of water can only exert the same pressure when moving with the same velocity, this will also be the same as in the first instance. But to impart to a greater mass an equal velocity to that of a smaller one, must necessarily absorb a greater amount of power; therefore the duration of the accelerated motion, as also the amount of accumulated power in the moving parts, and (in consequence of this, the duration of the retarded motion), will be considerably increased. From this we infer that, as the fly-wheel has no influence on the velocity of the mill during the equilibrium, it may be dispensed with in all machines experiencing a constant resistance at the working point, as only tending to increase the resistance due to friction upon the bearings.

43. (*d*). *The influence of the geometrical proportion of the moving parts.* We now remove part of the gearing of the mill, consisting of two spur-wheels of equal diameter, in contact with one another, and substitute two others, of which the one on the driving shaft is half the diameter of the one on the other shaft. Leaving all other conditions the same as in the first instance, the result of this change will be that, in order that the millstones shall move with the same velocity, the water-wheel must move with twice its previous velocity, and the pressure of the water upon the wheel will consequently be reduced to 1 cwt. The question arises, whether this alteration is beneficial to the action of the mill? Now, there are two conditions to be fulfilled for the mill to work with its maximum effect: (1), The most advantageous relative velocity of the floats of the water-wheel to that of the stream has to be considered. It has been found by expe-

rience, that the greatest amount of power is transmitted when the velocity of the former is one-third that of the latter. (2), There is a velocity beyond which the millstones cannot be driven without scorching the crushed grain by the heat resulting from friction. This velocity for ordinary sized stones should not be more than about 130 revolutions per minute. If, therefore, the mill, when under its former arrangement, was fulfilling these conditions, then it is clear that the above change will be detrimental to its action.

44. *Machines experiencing regular periodical equilibrium.* To represent this class, we will consider the action of a timber frame driven by a steam-engine.

45. (*a*). *Influence of the motive power.* Suppose that the pressure in the steam-cylinder is such, that when the timber frame is fully at work, the blades make 100 cuts per minute. There will consequently be 100 down strokes and 100 up strokes of the saw-frame in one minute, and the duration of the equilibrium will be = $\frac{1}{200}$th of a minute *plus* the difference between the time of duration of the working stroke and that of the returning stroke; for, as the pressure in the steam-cylinder is constant, and as, during the returning or up stroke of the saw-frame there is no other work done than the raising of the weight of the same, this will naturally take place at a quicker speed than the working stroke. If we now admit a greater pressure of steam in the cylinder, so that the saws, instead of making 100 cuts, now make 200 cuts per minute, the question arises, is this increase of speed of advantage or detrimental to the working of the timber frame? In making twice the number of cuts in the same time, the saws have to tear through twice the number of wood-fibres, and one would suppose that twice the power was required to do this. But this is not the case; for we know that the *vis viva* in a mass increases with the square of the velocity, and therefore the saw-frame in descending with twice the velocity that it did before acts with more than twice the force on the fibres. Here

then is the gain or saving of power with an increased velocity; but how is it with the up stroke? The heavy saw-frame has to be raised with twice the velocity that it had before; and on the same principle that we have a gain of power with increased velocity during the working stroke, we have a loss of power during the returning stroke; and the nature of the answer to our question as to the advantage of going at a quicker speed is dependent on the proportion of this gain, on the one hand, to the loss, on the other.

46. (*b*). *Influence of resistance.* Leaving the pressure in the steam-cylinder as it was at first, we will now put a balk on the carriage that is half the depth of the former one. The saws would now cut through only half as many fibres at one stroke as they did before, if the length of stroke and speed with which the timber was brought forward, remain unaltered; for during part of the down stroke the saws will not be cutting at all. To obviate this we must, therefore, move the balk twice as fast as we did before, when we shall have virtually the same resistance as in the first instance, and the speed of the machine will also remain the same. But if we have the same sized balk as at first, and bring it forward with half the speed, then, though the saws will be cutting during the whole of the down stroke, they will only be doing half the work, and the machine will therefore move at a greater speed.

An increase of speed does not result here in a decrease of pressure exerted by the motive power, from the same cause as with the water-wheel; for steam, as long as the velocity of the piston does not exceed 4000 feet a minute, will always exert a constant pressure at whatever speed the piston travels. The cause of a decrease in pressure exists in the fact that we have to generate the motive power; and are able to generate only a certain quantity of steam at a certain pressure in a given time, if we consume more than this quantity during this time, we must naturally decrease the pressure. The consumption of steam during the equilibrium will, therefore, be equal to the production of the same in the boiler.

47. (*c*). *Influence of the amount of mass in the moving parts.* We now introduce a heavy fly-wheel on the crank shaft. The result of this will be that, as during the up stroke of the frame there is no work done, the surplus power will be spent in accelerating the fly-wheel, and accumulating power in the same, which will be given off again during the down stroke. This will tend to equalise the motion of the machine, and approximate its action to that of the first class: the action of the fly-wheel here, therefore, is beneficial.

48. (*d*). *Influence of the geometrical proportions of the moving parts.* We remove the crank by means of which the saw-frame is moved up and down, and place in its stead another, in which the distance from the centre of the shaft to the centre of the crank-pin is twice what it was in the first. Leaving all else the same as it was in the first instance, the result of this change will be, that the length of stroke of the saws is twice what it originally was; and, consequently, if they are cutting during the whole of the stroke, the amount of work performed by them will be twice what it was at first. The pressure in the steam cylinder, however, remaining the same, the increase of work performed during one stroke will result in a decrease of speed; and, as we have before seen that, with a decrease of speed we have a more than a proportionate decrease of work performed by the same amount of motive power, it is evident that this alteration does not act beneficially on the working of the machine.

The length of stroke of the saw-frame should always be regulated by the size of timber it has to cut; it should be about equal to the depth of the balk.

49. *Machines experiencing irregular periodical equilibrium.* To represent this class we will consider the action of a rolling-mill driven by a water-wheel.

Here the work will take place in the following manner:—While the bar-iron is being heated in a furnace, the water is allowed to flow on to the water-wheel, by which it and all other moving parts, including a heavy fly-wheel, are soon set into rapid motion, as the only resistance for the motive power to overcome consists in the friction in the bearing and the inertia of the moving parts. During this time, therefore, a great amount of motive power is being accumulated in the form of momentum of inertia, particularly in the fly-wheel, which forms a kind of artificial motive power, exceeding in intensity, in many instances, by several times, that momentarily developed on the water-wheel. The bar of iron, at a welding heat, is now passed through the rollers, commencing at the largest grooves, and then consecutively through the smaller ones, until it has received the required scantling and length. The supply of water will now be cut off until another bar of iron is ready for operation.

50. The whole of the action during this time may be represented graphically by the figures 8 and 9, Plate I. The *abscissa* $A\ L$ and $A_1\ L_1$ represent the distance passed through by the circumference of the water-wheel during the whole of the operation. The ordinates of the curved line P represent the power with which the water acts, during this time, on the circumference of the wheel; the ordinates of the line R represent the resistance of the work reduced to the circumference of the water-wheel. The ordinates of the curve M represent the momentum of inertia contained in the moving parts during consecutive portions of the whole time. Lastly, the ordinates of V represent the velocity at different periods of the water-wheel.

The distance A B represents the accelerated motion of the machine, or the time from the water first setting the wheel in motion till the bar-iron is first introduced between the rollers.

B C, D E, F G, H I, are the distances passed through by the circumference of the water-wheel during the consecutive operations of rolling, while C D, E F, G H, I K, represent the same during the time that the heated iron is brought back over the rollers, again to be passed through the next groove; during these periods therefore the mill is not working. K is the period when the water supply is cut off, and K L, the duration of the retarded motion of the machine, which lasts until the whole momentum of inertia in the moving parts has been consumed by the friction in the same, the amount of which is represented by A a, B b, C c. The resistance while the iron is passing through the first, second, third and fourth grooves, is represented respectively by B b, D d, F f, H h; it occurs suddenly, decreases in intensity, and increases in its time of duration, as the bar-iron becomes of a smaller scantling and greater length. The power P, with which the water acts on the circumference of the wheel, changes with the velocity of the latter; as the slower the wheel goes the greater will be the amount of water that will flow on to it during a certain period. This power will therefore increase at first until the wheel has made about half a revolution, and will then decrease as the motion of the mill becomes accelerated; it will increase again slightly when the iron is first brought between the rollers, and retards the motion of the mill, and decrease again during the intervals when the work is off. Thus the pressure will fluctuate during the whole time of action, as shown by the line P.

As just stated, the curve V, representing the velocity of the wheel, will gradually increase from nothing during the acceleration; will drop when the work is on; will rise slightly when it is off, and will fluctuate thus in decreasing, as the accumulated momentum of inertia is consumed. The curve M, representing the momentum of inertia, will be similar in its form to V, only, as it varies with the square of the latter, the oscillations will become more exaggerated.

51. The action of the fly-wheel is here, as in general with all machines experiencing sudden changes in the intensity of the resistance, of the utmost importance. Indeed, it is in many cases the only means by which it is rendered possible, with only a small available quantity of motive power, to overcome the great resistance experienced in the operation of rolling iron. This is done, as before stated, by setting the mill in motion some time before the work is put on, thus expending the power in accelerating the fly-wheel and other moving parts, and concentrating, as it were, the power in them to such an extent as to make the motive power that is momentarily generated in the receiver, only equal to a fractional part of that accumulated. Were we to dispense with this feature, we should require, first, an amount of motive power of such magnitude as to overcome the greatest amount of resistance that can occur; secondly, in order that there should not be too great a variation in the speed of the mill, we should require the means of regulating momentarily the supply of motive power.

CHAPTER III.

OF THE USEFUL AND DETRIMENTAL EFFECTS OCCURRING IN THE ACTION OF MACHINERY IN MOTION.

52. If it were possible to construct a machine of an absolutely perfect nature as regards the reception and transmission of power, the effect received at the impelled point, as also that given off at the working point, would be equal to the total or absolute effect contained in the motive power. But as, in practice, this perfection is unattainable, the effect communicated to the receiver is less than the absolute effect contained in the motive power, and the effect given off at the working point is also less than that communicated to the receiver.

A correct valuation of the capabilities of a machine can therefore only be arrived at by a knowledge of the following :—

1. The absolute effect contained in the motive power.

2. The useful effect transmitted by the receiver.

3. The loss of effect through friction, vibrations, and inaccuracy in the construction of the working parts.

4. The useful effect developed by the working point or tool.

The absolute effect contained in the motive power may be ascertained, as already indicated in the chapter on motive power. The other effects may be ascertained by means of three methods: (1) valuation, (2) measurement, (3) calculation. Each method has its advantages, and will, under certain circumstances, be preferred to the two others.

53. The method of valuation affords the least accuracy, and can only be employed where the engineer has had considerable experience and practice in the estimation of these effects. Where this is the case, it is however, through its immediate nature, of great value.

This method consists in estimating, by means of comparison with results arrived at in parallel cases, the average useful effect that may be derived from a certain motor when acting upon a particular description of receiver or engine. For instance, the average value of the useful effect derived from water power has been found to be :—

1. For undershot water-wheels . .	0·3 to 0·4 of the total power.
2. For breast-wheels . . .	0·5 " 0·6 ,, ,,
3. For overshot water-wheels . .	0·6 " 0·7 ,, ,,
4. For turbines (horizontal water-wheels)	0·6 " 0·7 ,, ,,

The average value of the useful effect derived from steam power, expressed by the weight of coal consumed per horse power in an hour, is, for different modes of employing it, as follows :—

Steam of 16 lbs. per square inch (12 lbs. vacuum) total pressure in cylinder, not expanded but condensed, giving an average pressure of 16 lbs. per square inch	8 lbs. of coal per H.P. per hour.
Ditto at 19 lbs. per square inch (12 lbs. vacuum) total pressure, expanded during ⅓ of the stroke and condensed, giving an average pressure of 16 lbs. per square inch (see diagram of expansion of steam)	7 lbs. ,, ,, ,,
Ditto at 22 lbs. total pressure, expanded during ½ the stroke and condensed, giving an average pressure of 16 lbs. . . .	6¼ lbs. ,, ,, ,,

Ditto at 27 lbs. total pressure, expanded during $\frac{2}{3}$ of the stroke and condensed, giving an average pressure of 16 lbs. . .	$5\frac{1}{2}$ lbs. of coal per H.P. per hour.
Ditto at 32 lbs. total pressure, expanded during $\frac{3}{4}$ of the stroke and condensed, giving an average pressure of 16 lbs. per square inch	$4\frac{3}{4}$ lbs ,, ,, ,,
Ditto at 37 lbs. total pressure, expanded during $\frac{4}{5}$ of the stroke and condensed, giving an average pressure of 16 lbs. . .	4 lbs. ,, ,, ,,
Ditto at 42 lbs. total pressure, expanded during $\frac{5}{6}$ of the stroke and condensed, giving an average pressure of 16 lbs. . .	$3\frac{1}{4}$ lbs. ,, ,, ,,

With these data, then, derived from experience, we shall be enabled to ascertain with tolerable accuracy what amount of useful effect may be gained by the employment of a certain description of motive power and a particular class of engine.

54. *The method of measurement* affords very valuable results if carefully carried out, and is particularly of service where it is required to ascertain whether a machine really affords the useful effect it was intended to give.

This measurement is carried out by means of a class of instruments termed *dynamometers*, with the most important one of which, namely, the steam-engine indicator, we have already made acquaintance. Another description of dynamometer is the *friction brake*, of which fig. 10, Plate I., is the most simple form, as invented by *Prony*. A represents a shaft, the power transmitted by which it is required to ascertain; for this purpose it is first disconnected with that part of the machinery to which it communicates the power, being left in connection with the part from which it receives the same. The drum B is then fixed concentrically upon it, the diameter of which is from 4 to 6 times, and whose width is from 1 to $1\frac{1}{2}$ times, the diameter of the shaft.

Round its circumference, which is accurately turned, are fitted the blocks of wood C and D, held in position by the bolts E E passing through them, as also through the lever F. These bolts, by being screwed tight, cause a certain amount of friction to take place between the blocks of wood and the circumference of the drum when the latter is set in motion, the intensity of which friction, at a certain number of revolutions of the shaft, it is the object of the instrument to measure. For this purpose the lever F is fitted with a receptacle for weights at G, the weight of which, as also that of the lever itself, is balanced by the counterpoise H, so that when the drum is at rest and there are no weights at G, the lever will remain in a perfectly horizontal position. I I are two fixed studs which prevent the lever from moving much beyond this position. To determine the amount of effect transmitted by the shaft A, we now proceed in the following manner :—

The shaft is now set in motion by the prime mover, the screws are then gradually tightened to such a degree that the shaft moves exactly with the velocity at which its useful effect has to be determined. Were it not for the stud I, the lever, with the whole brake apparatus, would move round in a circle with the shaft, so that this stud has to withstand a certain force from the lever pressing against it. We now place very carefully and gradually upon G such a weight as will counteract the pressure of the lever against I, and bring it back to its horizontal position. It is evident that this weight in combination with the lever F, expresses the quantity of friction between the blocks and the drum, as it holds it *in equilibrio*. Therefore if G is the weight on the scale, R the radius of the drum, and L the length of the lever from the centre of the shaft, the quantity of friction will be expressed by—

$$G \frac{L}{R} \quad . \quad . \quad . \quad I.$$

If, in addition, we count the number of revolutions $= n$

per minute of the shaft, we find the linear velocity of the circumference of the drum per second:

$$v = \frac{2\,R \times 3{\cdot}14 \times n}{60}.$$

As the above expression of the resistance from friction is overcome by this velocity, v,

$$G\,\frac{L}{R}v \quad . \quad . \quad . \quad \text{II.}$$

expresses the useful effect developed by the shaft A.

The expression $\frac{L}{R}\,v$ also represents the velocity V with which the end of the lever F would move, were it fixed to the drum B and revolved with it; the useful effect may therefore also be expressed by

$$G\,V$$

Or, *the useful effect developed by the shaft is equal to the product of the weight placed upon the scale, multiplied with the velocity with which the same would be caused to revolve, were it fixed to the shaft.*

The success of experiments conducted with this apparatus is dependent upon one or two things which we will note here. The drum should not be of too small a diameter, else the intensity of the necessary amount of friction would be such as to cause heating and even burning of the wood-blocks. The drum should moreover be fixed very accurately concentrical with the shaft, and its circumference be turned very smooth. During the motion, soap and water should constantly be poured upon the rubbing parts, both to prevent heating and to give an even motion. For this purpose it is also as well that, before the experiments are made, the rubbing surfaces of the wood-blocks should be allowed to become slightly charred, which renders them of a very smooth and velvety texture. The same apparatus may be adapted to vertical shafts by fixing the scale G with the weights to a cord, passing the same over a fixed pulley, and attaching the other end to the lever F.

55. *The method of calculation* can only be applied where the laws are accurately known according to which power and resistance act, but in such cases affords a very detailed insight into the manner in which every element of construction influences the action of the machine.

It consists in ascertaining, firstly, the absolute effect imparted either by the motor to the receiver, or by one part of a machine to the next in connection with it; secondly, in determining the different *partial effects*, which cause a diminution of the useful effect, and depend in their nature upon the particular class of engine or machine under consideration.

The sum of the latter deducted from the former will then give the useful effect to be gained from the employment of a particular class of machine. Thus, for instance, if we want to ascertain the gross power requisite for working a stamping mill, we have to calculate: 1, the useful effect requisite for raising the stampers; 2, the loss of effect resulting from the sudden impact of the cams in raising the stampers; 3, the loss of effect from the acceleration of the mass of the stampers in their upward motion; 4, the loss of effect through the friction of the cams against the studs of the stampers, and also of these latter in their guides; 5, the loss of effect through friction in the bearings of the cam shaft.

As another instance, we will consider the power requisite for pumping water to a certain height, besides conducting it to a considerable distance. This will be subdivided into: 1, the useful effect requisite for raising the water; 2, the loss of effect resulting from the acceleration of the mass of the moving water; 3, the effect necessary to overcome the friction of the water flowing through the conduit pipes; 4, the loss of effect caused by the friction in the pumps and other moving parts; 5, the loss of effect resulting from concussions and sudden changes of velocity in the mass of water, owing to the opening and closing of valves and alteration of sectional area.

We will now ascertain by what means these partial effects may be computed.

56. *Loss of effect through friction.*—To ascertain the power requisite to overcome a certain amount of friction, it is firstly necessary that we should know the intensity of the pressure exerted on the rubbing parts. This being done, then the product of this pressure, with the co-efficient of friction corresponding with the nature of the substances in contact, will give the amount of resistance caused by the friction, and this again, multiplied with the relative velocity of the parts in contact, will give the power, expressed in *pounds feet,* absorbed by friction.

Where we have to deal with a train of mechanism in which we know from experience that the total amount of power absorbed by friction in all the moving parts, forms but a fractional part of the transmitted power, we shall form an approximate estimate of the same, sufficiently accurate for practical purposes, if we ascertain by the above method the intensity of friction in the first moving part, and then for each other link in the train simply multiply this amount with the respective velocity ratio to the first link.

57. The laws which regulate the resistance from friction have been discovered by Colomb, and latterly by Morin, through the aid of numerous experiments. The results obtained by them are mainly as follows:—

1. The resistance is dependent on the nature of the substances and the state of the rubbing surfaces. The friction of wood on wood is less when the grain is at right angles than when it is parallel in both pieces. Moisture and oil increase; tallow, soap, and graphite decrease the same. For metals, oil and lard decrease the amount of friction.

2. The resistance is greater at the commencement of motion after a rest of some duration, than when the motion has continued for some time.

3. The resistance is less with bodies experiencing a circular motion, as for instance the bearings of shafting, than with those experiencing rectilinear motion.

4. The resistance is independent of the amount of surface in contact, as long as the same is large enough to prevent heating or abrasion.

5. The resistance is independent of the velocity of the rubbing parts, as long as no heating or abrasion takes place.

6. The resistance is in direct proportion to the pressure exerted upon the rubbing parts, within the above-named limits.

Let f represent the resistance from friction between two surfaces, resulting from a pressure of 1 lb. F the same due to a pressure of P pounds, and v the velocity with which the bodies move, then the effect E expressed in *pounds feet* necessary for overcoming this friction will be

$$E = P f v, \text{ and}$$
$$F = P f.$$

58. The different values for f, called the *co-efficient of friction*, for different materials under different circumstances according to the results arrived at by Colomb and Morin, are subjoined.

TABLE I.

CO-EFFICIENT OF FRICTION AT THE COMMENCEMENT OF MOTION.

Description of Surfaces in Contact.	Direction of the Grain.	State of Surfaces.	Co-efficient of friction.
Oak upon oak	parallel	without unguent	0·62
	,,	with dry soap	0·44
	at right angles	without unguent	0·54
	,,	with water	0·71
Oak upon elm	parallel	without unguent	0·38
Elm upon oak	,,	,,	0·69
	,,	with dry soap	0·41
	at right angles	without unguent	0·57
Ash, fir, beech, upon oak	parallel	,,	0·53
Tanned leather upon oak	flat	,,	0·61
	leather upon edge	,,	0·4[illegible]
		moistened with water	0·79
Black dressed leather upon a plane surface of oak	parallel	without unguent	0·74
Black dressed leather upon a rounded surface of oak	at right angles	,,	0·47
Hemp cords upon oak	parallel	,,	0·80
Wrought iron upon oak	,,	with water	0·65
		without unguent	0·62
Cast iron upon oak	,,	,,	0·62
		with water	0·65
Gun metal upon oak	,,	without unguent	0·62
Ox-hide as a piston packing upon cast iron	flat or on edge	with water	0·62
		with oil, soap, or lard	0·12
Leather straps on cast iron drums	flat	without unguent	0·28
		with water	0·38
Cast iron upon cast iron	,,	without unguent	0·16
Wrought iron upon cast iron	,,	,,	0·19
Oak, elm, beech, iron, cast iron, and bronze, two and two in contact	,,	with tallow	0·10
		with oil or lard	0·15

TABLE II.

Co-efficient of Friction after the Motion has continued for some time.

Description of Surfaces in Contact.	Direction of Grain.	State of Surfaces.	Co-efficient of Friction.
Oak ak	parallel	without unguent	0·48
	,,	with dry soap	0·16
	perpendicular	without unguent	0·34
	,,	with water	0·25
	end grain upon grain lengthways	without unguent	0·19
Elm upon oak	parallel	,,	0·43
	perpendicular	,,	0·45
Ash, fir, beech, wild pear, upon oak	,,	,,	0·36–0·49
Wrought iron upon oak	,,	,,	0·62
	,,	with water	0·26
	,,	with dry soap	0·21
Cast iron upon oak	,,	without unguent	0·49
	,,	with water	0·22
	,,	with dry soap	0·19
Gun metal upon oak	,,	without unguent	0·62
Leather belts upon oak	,,	,,	0·27
Tanned leather upon oak	flat or on edge	,,	0·30–0·35
		with water	0·29
Tanned leather upon cast iron or bronze	,,	without unguent	0·56
		with water	0·36
		greased and with water	0·23
		with oil	0·15
Hempen ropes upon oak	parallel	without unguent	0·52
	perpendicular	with water	0·33
Oak and elm upon cast iron	parallel	without unguent	0·38
Wrought iron upon wrought iron	,,	,,	*
Wrought iron upon cast iron and gun metal	,,	,,	0·18†
Cast iron upon cast iron and gun metal	,,	,,	0·15†
Gun metal upon Gun metal	,,	,,	0·20
Gun metal upon cast iron	,,	,,	0·22
Gun metal upon wrought iron	,,	,,	0·16†
Oak, elm, beech, cast iron, wrought iron, steel, and gun metal, one upon the other or itself	,,	with tallow, lard, oil, or grease	0·07 to 0·08
		slightly greasy	0·15

* Abrasion ensues. † Slightly greasy.

TABLE III.

FRICTION OF GUDGEONS AND SHAFTING IN THEIR BEARINGS WHILE REVOLVING.

Surfaces in Contact.	State of Surfaces.	Co-efficient when arranged, in the usual way.	Co-efficient when arranged, continuously
Cast iron axles in cast iron bearings . . .	coated with olive oil, lard, tallow, or soft gome	0·07–0·08	0·054
	with ditto, wet .	0·08	
	with asphaltum . .	0·054	
	greasy . . .	0·14	
	do. wet . . .	0·14	
Ditto in gun metal bearings	with olive oil, lard, or soft gome . .	0·07–0·08	0·054
	greasy . . .	0·16	
	do. and wet . .	0·16	
	very slightly greasy .	0·19	
Ditto in lignum vitæ bearings	with oil or lard . .	0·10	0·090
	without unguent .	0·18	
	with lard and molybdenum . . .	0·14	
Wrought iron axles in cast iron bearings . . .	with olive oil, tallow, lard, or soft gome	0·07–0·08	0·054
Ditto in gun metal bearings	with olive oil, lard, or tallow . . .	0·07–0·08	0·054
	with hard gome . .	0·09	
	greasy and wet . .	0·19	
	slightly greasy . .	0·25	
Ditto in bearings of lignum vitæ	with olive oil, or lard .	0·11	
	greasy	0·19	
Gun metal axles in gun metal bearings . .	with oil . . .	0·10	
	with lard . . .	0·09	
Ditto in cast iron bearings	with oil or tallow .	–	0·045 to 0·052
Lignum vitæ axles in cast iron bearings . . .	with lard . . .	0·12	
	greasy . . .	0·15	
Ditto in bearings of lignum vitæ	with lard . . .	–	0·07

59. Friction, and the vibration ensuing therefrom, forms the principal antagonistic element opposed to the economical and accurate working of machinery. Its utmost possible reduction should be a subject of the greatest importance with the engineer.

The principal cause of friction lies doubtless with the absence of absolute smoothness of the rubbing surfaces, the inequalities in the one griping into those of the other, similar to the action of cog-wheels, thus making it necessary, in order that motion can ensue, that the protruding particles in the one surface should be crushed, bent, or broken off by those of the other; herein the cause for abrasion and vibrations. The object of lubrication is, therefore, to fill up the small cavities with a substance for which the materials of which the surfaces consist have no hyproscopic affinity; that is, into which it does not penetrate, thus preventing the inequalities in the two surfaces from having so great a hold upon one another, and decreasing the friction accordingly. If, on the other hand, the substance employed for lubrication penetrates into the pores of the surfaces, then the molecular attraction between the particles of this substance will cause a closer contact of the rubbing surfaces, and therefore also a firmer griping to take place between the inequalities of the same, thus increasing the friction and abrasion instead of diminishing the same.

60. Of still greater importance than the loss of power through friction are the other effects of the same; namely, abrasion, vibration, and heating of the moving parts, as these may cause the eventual destruction of a machine. The vibrations caused by friction may be greatly increased from other dynamical influences, and these vibrations being transmitted through the different parts, escape through such as offer a direct communication with the earth. Now, unfortunately, these parts are generally just the very ones upon which their action is most detrimental, namely, the bearings of shafting, &c. It is therefore of the greatest importance that these should not only be constructed and secured to their foundations with the greatest possible care, but also that every means should be employed, such as careful lubrication, to reduce this destructive action to a minimum. Perhaps nowhere is this more necessary than

with the gudgeons of turbines, where about 10 per cent. of the absolute effect is absorbed by vibrations caused by the impact of the water. These vibrations have all to escape through the one small gudgeon upon which the turbine runs, and which besides has to bear the whole pressure caused by the weight of the turbine and shaft, as also of the column of water. An appropriate construction for these will be indicated elsewhere.

61. To prevent vibrations as much as possible, it is necessary that we should employ materials and methods of construction affording the greatest possible rigidity; that all parts at rest, as also those experiencing rotatory motion, in which the centre of rotation is also the centre of gravity should be as massive as circumstances will allow; but that all parts experiencing an oscillating or reciprocating motion, or a rotatory motion in which the centre of rotation is *not* the centre of gravity, should be made as light as possible, and their action as much as possible balanced. By this is meant that a balance weight should be brought in such a manner in connection with the reciprocating part, that by the motion of the two the centre of gravity of the whole machine shall remain undisturbed. This is highly necessary where the reciprocating parts must unavoidably be of a certain strength and move with great rapidity, for, the centre of gravity of a machine always striving to remain stationary, the whole machine would, if suspended so as to give it liberty of action, move to a certain extent in an opposite direction to the reciprocating parts within it, thus making it evident that where the machine is fixed and cannot follow these impulses, violent strains and consequent vibrations must ensue. The reader may easily convince himself of the intensity of the power tending to move the machine when the reciprocating parts are unbalanced, by constructing a simple model as shown in Fig. 11, Plate I., representing the framework, wheels, and reciprocating parts of a locomotive engine

suspended so as to have to a certain degree liberty of action. If the weights W representing the piston, piston-rod, connecting rod, crank, &c., of the locomotive, are made equal to about 3 lbs. each, with a stroke of about 4 inches, and they are set in motion with considerable velocity by means of the pulley *a* and belt *b*, it will be found beyond the power of one man to prevent the whole frame from moving about, in striving to compensate for the displacement of the centre of gravity caused by the reciprocating weights. In actual locomotives this disturbing influence was so unpleasantly evident in the motion of the train that it has been for some time the practice of engineers to fix *balance weights* in the driving wheels to counteract it.

62. There is yet another cause for the avoidance of heavy reciprocating parts, which, although not belonging to the subject under consideration, may as well be mentioned here, namely, at the commencement of each stroke a certain velocity has to be communicated to the mass, which therefore absorbes a certain amount of motive power in the form of vis viva. At the end of the stroke an opposing force has to be employed, not only to balance this effect in order to bring the mass to a standstill, but also to impart to the same an equal velocity in the reverse sense. Thus then, at every stroke there is a loss of effect equal to twice that absorbed by the reciprocating parts, and this, when the velocity is great, is very considerable, as its value is expressed by $\frac{W \times v^2}{2g}$ where $W =$ the weight of the reciprocating parts, $v =$ their velocity and $g = 32{\cdot}16$ feet.

CHAPTER IV.

OF THE CONTRIVANCES FOR REGULATING THE MOTION OF MACHINES.

63. In all mechanical processes there is a cer ain velocity both of the impelled and working point, at which the work is performed to the greatest advantage. These velocities being ascertained either by practical experience or by means of calculation, we have so to proportion the moving parts intervening between the receiver of power and the working point, that when this most advantageous velocity occurs in the former, it of a necessity causes the same to take place in the latter. This being effected then, no matter what the amount of impelling force at the one point, no matter how great the resistance at the other, the proportion between these two velocities will always remain the same; so that an increase or decrease of the one will always cause a proportionate increase or decrease of the other.

As a deviation from this most advantageous velocity—particularly of the working point—in many cases is very injurious to the quality of the substance manufactured; and as, further, in most mechanical processes it is almost impossible to keep up a constant resistance at the working point, so as at every instant to balance the action of the impelling power, or to keep up what we have termed a constant dynamical equilibrium, owing, in works of any magnitude, to the fact that there is always some working machine or other thrown out of gear for the moment thus causing a decrease of resistance and a resulting increase of speed in the other machines; as, we say, such casualties are inevitable, it becomes a subject for the consideration of the engineer to provide such contrivances in connection with the machinery as will *govern* the motion produced to such a degree as to reduce these fluctuations in its velocity to a minimum.

Such contrivances are known under the terms *regulators* or *governors* of motion, and may be divided, according to the principle upon which they act, into three classes, namely:

1. Such as absorb the surplus power resulting from a sudden decrease of resistance, and give it out again when the resistance is increased. To this class belong the *fly wheel* and *balance weights*.

2. Such as, in receiving an increased velocity from a decrease of the resistance, will, owing to the nature of their contrivance, cause a decrease in the amount of motive power that momentarily acts upon the receiver, and *vice versa*. To this class belongs the so-called *governor*, or conical pendulum.

3. Such as will, when an increased velocity is the result of a decrease of the useful resistance, create an artificial resistance, upon the overcoming of which the surplus power is uselessly expended; to this class belong the *friction brake*.

64. The *fly wheel* is generally of a large diameter, to give a great circumferential velocity to the massive metal rim, without having to drive it at an excessive speed. Its object, as before stated, is to absorb any surplus of motive power in the form of momentum of inertia, and to give the same off again when there is a deficiency of power, without the velocity of the same and of the rest of the machinery undergoing any very perceptible change.

If we call M the mass of the fly-wheel rim, D the difference of the effects produced by the motive power, and consumed by the resistance during a certain interval of time; v and V the velocities of the fly-wheel rim, at the commencement and at the end of this interval of time, then supposing that the momentum of the other moving parts is small as compared with that of the fly-wheel, we have, approximatively:

$$M (V^2 - v^2) = D$$

From which,

$$V - v = v \left\{ \sqrt{1 + \frac{D}{M v^2}} - 1 \right\}$$

But as, generally, the value of D is small as compared to that of M v^2, we may suppose approximatively:

$$\sqrt{1 + \frac{D}{M v^2}} = 1 + \tfrac{1}{2} \frac{D}{M v^2}$$

and then we have,

$$V - v = \tfrac{1}{2} \frac{D}{M v}$$

From which we deduce that the variation in the velocity of the fly-wheel is directly proportional to the differential effect, D, absorbed by the same, and inversely proportional to the mass and velocity of the rim. Therefore a heavy fly-wheel, moving at a considerable speed, is capable of governing the motion of a machine to such an extent, that even where the variation of pressure on the receiver, and of the resistance at the working point, is very considerable, no very great alteration in the velocity of the machine can take place. The necessary weight for the fly-wheel rim can only be ascertained accurately by calculation, when the law according to which a periodically occurring fluctuation of the resistance takes place, is accurately known; but even where this is not the case, we may state as a rule that the momentum corresponding to the mean velocity of the fly-wheel should be proportional to the useful effect developed by the machine. If we call W the weight of the fly-wheel rim; V the mean velocity of the same; N the horse power of the machine; α a constant, to be determined by experience, depending on the nature of the machine and the degree of uniformity in its motion that may be required; then we have

$$\frac{W}{2g}V^2 \quad 550\,\alpha\,N,$$

from which

$$W = \frac{2g\,550\,\alpha\,N}{V^2},$$

The value of α varies from 30 to 60; that is to say, the mass of the fly-wheel rim should be such that the momentum corresponding to the mean velocity is equal to the useful effect developed by the motive power during from 30 to 60 seconds.

The fly-wheel is employed with advantage, as already stated: 1stly, in all machines having a rotatory motion, which experience a regular or irregular periodical fluctuation in the resistance at the working point; 2ndly, to render the unequal motion resulting from the employment of the crank more uniform.

65. The employment of *balance weights* is of advantage: 1stly, when the centre of gravity of certain moving parts alternately moves up and down; 2ndly, when a machine has to lift certain weights with a sudden motion; 3rdly, when resistances have to be overcome that vary very considerably in intensity, but which variations recur regularly. They should be of such a size that, by means of the same, the motion of the machine occurs with the greatest possible uniformity.

In a timber frame, for instance, to equalise the motion, we should have to balance: 1stly, the unequal action of the crank; 2ndly, the weight of the crank, the connecting rod, the saw-frame and saws; 3rdly, the unequal resistance that occurs on account of the saws only working during the down stroke; 4thly, the periodical resistance due to the moving forward of the timber carriage. All these inequalities are best balanced by means of a fly-wheel, to the rim of which a balance weight is attached. In direct-acting pumping engines, in which the steam works expansively in one cylinder, a balance weight is necessary to absorb part of the power developed by the steam during the first part of the stroke, and to give it off again during the latter part, when the pressure of the steam is low and, unaided, not equal to the constant resistance due to the pump rods. Mr. E. A. Cowper proposes to employ such a balance in the manner shown in Fig. 12, Plate I., where the same is attached to the top of the beam of the engine in such a manner, that it has to be lifted during the first half of the stroke, and assists the engine with its descending weight during the latter half.

66. The *conical pendulum* or *governor* as shown in fig. 13, Plate I., regulates the supply of motive power in the following manner. The time of rotation determines the height *l* from the point of intersection of the continuation of the arms *a* to the centre of the balls *b*, or in other words, the greater the number of revolutions per minute the more the balls will fly out, thus causing, by means of the links *c*, the rod *d* inside the tube *e* to rise and fall according as these balls rise or fall. The rod *d* is connected with the throttle valve in the steam pipe by means of levers and connecting rods, and therefore, when the engine is going too fast, owing to a decrease in the resistance, the balls will fly out, cause the rod *d* to rise, and thus shut off, more or less, the supply of steam. If, on the other hand, the engine moves too slowly on account of extra work being put on, the balls fall and thereby actuate the opening of the valve. The same contrivance, with slight modification, may be made to open or shut the sluice of a water-wheel.

The conical pendulum performs one rotation during the time that a common pendulum of a length equal to the vertical height *l* makes two oscillations; therefore if the time of oscillation of a simple pendulum is represented by

$$t = \pi \sqrt{\frac{l}{g}} = 0{\cdot}5537 \sqrt{l}.$$

π being $= 3{\cdot}1416$, $g = 32{\cdot}190$, $t =$ time in seconds, and $l =$ the length of the pendulum in feet, then the time of rotation of a conical pendulum of a vertical height $= l$ will be

$$t' = 2\pi \sqrt{\frac{l}{g}} = 1{\cdot}1074 \sqrt{l},$$

or the number of revolutions per minute of a given height l in feet:

$$n = \frac{60}{1{\cdot}1074 \sqrt{l}} = \frac{54{\cdot}18}{\sqrt{l}},$$

or, for the value of l in inches: for the common pendulum

$$n = \frac{60 \times \sqrt{12}}{0{\cdot}5537 \sqrt{l}} = \frac{375{\cdot}36}{\sqrt{l}};$$

for the rotatory pendulum

$$n = \frac{54{\cdot}18 \times \sqrt{12}}{\sqrt{l}} = \frac{187{\cdot}68}{\sqrt{l}}.$$

The length of the seconds pendulum for the latitude of London (51° 31′ 34″ N.) is, according to the above, 39·139 inches, therefore a conical pendulum of the same height will make 30 revolutions per minute.

For two conical pendula of the length l and l' and respective time of rotation t and t', we have the proportion

$$t : t' = 1{\cdot}1074 \sqrt{l} : 1{\cdot}1074 \sqrt{l},$$
$$t^2 : t'^2 = l : l'$$

That is: *The lengths of pendula are as the squares of their periods of rotation*; or, *their periods of rotation are as the square roots out of their length.* The defect of the rotatory pendulum, when employed as a governor or regulator of power, lies in its over sensitiveness: the least alteration in speed causes the balls to fly out or fall, and the momentum acquired by the motion carrying them beyond the mark, they always shut or open the valve too much, thus causing a continual fluctuation to a certain extent in the speed of the machinery. To avoid this as much as possible, the balls should be made of considerable size, and the mean speed as slow as convenient, care of course being taken that the extent of motion of which the arms are capable above and below their normal position, shall be equal. The angle of the arms with the perpendicular, when in their normal position, should be about 30°, for the governor to act to the greatest advantage.

67. Regulators of motion whose action consists in consuming the surplus power without useful effect, should obviously not be employed in machinery where economy is an object. They are employed however: 1stly, where the power they consume is of no importance, as for instance the fly regulator in the common pendulum clock, which causes a resistance by the vanes striking the air with a variable speed. 2ndly, Where the motion produced is an object, but the power expended none; as for instance the brake attached to crabs and cranes, employed for lowering heavy weights, to prevent these obtaining too great a velocity due to their acceleration.

CHAPTER V.

ON THE GENERAL PRINCIPLES OF CONSTRUCTION OF MACHINERY.

68. The construction of a machine, as regards the description of mechanical contrivances contained in the same, is mainly governed by the species of work that it is destined to perform. We have already elsewhere classified machines in general under two principal heads, namely: 1, Receivers of power, as engines, &c.; 2, Working machines. The consideration of the first class, as regards their principles of construction, is comparatively of a simple nature to that involved in the second class. In the first instance we have to consider the laws according to which the motive power acts, and to form our engine in such a manner that it should take up a maximum amount of the total effect developed by the same; the principal feature of the engine consisting of certain parts upon which the motive power impinges directly, and forces into motion, either rotatory round a fixed axis, or reciprocating rectilinear, which motion it is the further duty of the engine to transmit in the most advantageous manner to the working machine.

In the second instance we have generally to deal with infinitely more intricate considerations, necessitating a knowledge of the construction of the raw material and that of the product to be manufactured from the same; involving, besides, the invention or the application of mechanical contrivances, often of the most subtle nature, to bring about the required change in these substances.

In general, the working machines may be divided, as regards the duties performed by them, into three classes. 1, Such as have merely to bring about an alteration in position of the material. 2, Such as have to operate a division, a mixing, or sorting, a compression or rarification of the material. 3, Such as operate a total or partial change in the structure of the raw material.

69. All raw material possesses certain mechanical and chemical properties dependent on its molecular structure, that is, on the manner in which the molecules are grouped together, and on the forces by which they are held in a state of equilibrium. According to this molecular structure, substances are hard or soft, brittle or tenacious, solid or fluid, crystalline or homogeneous. In organic bodies the mechanical properties are generally founded on their cellular structure.

In addition to this physical structure of the raw material we have, in arranging our mechanical process, to take into consideration the state of impurity in which the substance is brought to market; this often necessitates certain preparing or cleansing processes, which have to precede the main manufacturing process, which again in many cases is divided into several partial processes, each operating a certain change in the raw material and advancing its manufacture a step towards completion. The nature of the manufactured product may differ either slightly or very much from the raw material. In the first case the process of manufacture consists merely of a compression, expansion, division, or aggregation of the mass; in the second, each separate particle undergoes a change as regards its relative position to the other particles, or else the raw material, consisting of a conglomeration of molecules differently constituted as regards their chemical or mechanical nature, becomes subdivided and separated according to these differently constituted particles; this latter process also often takes place in the reverse order, *i. e.* a variety of differently constituted materials are incorporated into one compound. Thus we have, as an illustration of the first method of manufacture, the compression by powerful agency of various substances to form a solid mass. Illustrative of the first-named of the second methods, we have the manufacture of cotton, consisting mainly in the arrangement of an entangled mass of fibres, so that they become straightened out and parallel to each other. Illustrative of the second-named, we have the grinding of corn, and the assortment of the products obtained into bran and flours of different qualities. Illustrative of the third process mentioned, we have the manufacture of gunpowder, where the constituents, charcoal, sulphur and saltpetre, are incorporated to form the inflammable compound so different in its nature from each of the component parts.

From this we see how essentially necessary it is, before we can venture upon the construction of machinery for executing these different processes, that we should make ourselves thoroughly acquainted with the nature, not only of the material in its unmanufactured state, but also of the manufactured product. Having acquired this knowledge, our next step is to set about the contrivance of machinery for performing the different operations required, and in doing so we must not lose sight of the conditions which a machine must fulfil, if its adaptation to the process of manufacture shall prove in every sense advantageous.

1. The product obtained by the machine must be of an unvariable quality, that is, the machine must be unerring in its operations.

2. The construction of the machine should be such that it does not require too great an amount of skill of the attendant. It would obviously be a fallacy to construct a machine for carrying out a certain operation, which machine should involve a greater expense in the remuneration of a sufficiently skilled attendant, than would be incurred were

the same quantity and quality of product to be produced by manual labour. This is not of likely occurrence ; but in our present age of machinery-mania, it would be well to bear in mind that there is one thing in which the most ingenious machine is wanting, and that is *human intelligence*, and that wherever this is constantly put in requisition to carry out a certain process, no machinery alone will be of avail.

3. The construction of the moving parts should be such as to reduce the amount of wear and tear to a minimum, and to render re-adjustment or reparation as easy as possible.

In contemplating the manufacture of a certain article, it may at first sight appear a desideratum to perform the whole operation in one machine. This, however, cannot be too carefully considered. It is true that the same reason for subdivision of labour, in the case of manual labour (viz.: the more the action of the mechanic is confined to one operation, the greater will be the perfection he will attain in it), will not hold good where machinery is employed. But still there is a very similar reason which would argue in favour of subdivision of labour also in machinery, and that is, that the simpler the operation and consequently the construction of a machine is, the more certain will be the result obtained from its operation. Here again, however, we may go too far, for in constructing a separate machine to perform every fractional portion of the whole process of manufacture, three questions are involved, namely, whether the advantage thereby gained overbalances ; 1, the increased expenditure incurred by the greater number of machines ; 2, the increased space of building which these involve ; and 3, the increased expense of transport consequent on the former. These are points for which a general rule cannot be given, but which will have to be determined according to the specialities of the process of manufacture in contemplation.

We have elsewhere described the working machine as "acting upon the substance to be manufactured in such a definite and invariable manner as to admit of no other result being obtained than the desired one." We will now shortly consider by what means we are enabled to construct machinery which shall satisfy these requirements.

70. It is clear that the elementary forms of a machine must owe their existence in the first instance to human intelligence and manual skill. Nature does not afford us directly any of those forms or principles to which a machine owes its efficacy; they entirely originate in the human mind, and are called into existence by the united effort of the mind and the hand.

Until within not very many years ago, this effort of mind and hand, exclusive, of course, of that involved in the actual invention, had to be re-exerted upon every reproduction of one and the same mechanical contrivance. This was effected by means of very great expenditure both of time and money, and after all, the machine when complete, did not in most cases work with that accuracy necessary to repay the great outlay on its construction. It was, therefore, not until mechanical contrivances, produced with the utmost possible accuracy and skill of which the human eye and hand was capable, were substituted for manual labour in the formation of machines intended for carrying out manufacturing processes, that the science of machinery received that impetus which, in the space of a very few years, has carried it onward towards the climax of perfection.

In constructing any mechanical contrivance which itself is again destined to be auxiliary in the construction of other machinery, we imbue that contrivance with the capability of reproducing manifoldly, in a comparatively short space of time, all the skill which has been expended in its own construction. This faculty of reproduction is termed the *copying principle*, and is contained in a greater or lesser degree in almost every machine intended for manufacture.

In many machines of the present day, this principle is carried out to its fullest extent; that is to say, the machine contains in itself an exact facsimile of the object it is intended to manufacture, and it actually is made by means of certain contrivances to copy the exact form of this model. Machines of this description are principally of American invention; they have been employed with very great success in the formation of many articles employed for war purposes, among which perhaps the most remarkable is the gunstock, which excentrically and intricately shaped piece of workmanship is now formed entirely by means of these highly ingenious machines. Another great advantage that this class of machine affords, beside their dispensing entirely with skilled manual labour, is, that where the article to be manufactured consists of several distinct parts, which have afterwards to be put together, any of these parts taken at random will fit together with an accuracy truly marvellous.

The application of these machines is confined to the process of moulding or shaping materials; but, as before stated, every working machine contains the same principle more or less; the tool, or working point, always moving in a prescribed manner, and with a predetermined velocity relatively to the material, or *vice versâ*, this motion being brought about by the adaptation of certain mechanical contrivances, which modify according to their nature the uniform rotatory motion of the shaft which supplies the power. The invention of these mechanical contrivances constitutes the principal feature in the construction of a machine. Frequently we shall find several solutions to one and the same problem, in which case we shall have to determine which of these solutions will afford the greatest working effect, or be the least expensive. In general, preference should always be given to such contrivances as will bring about the desired motion with the minimum amount of moving parts, as entailing the least loss of power, and the greatest accuracy in the motion; it being a popular error to consider those contrivances as most ingenious that involve the most intricate mechanism in their construction. Yet we must guard against seeming simplifications, which, though consisting of less moving parts, involve a greater loss of power. Such, for instance, is the application of an endless screw and worm-wheel in the place of a whole train of wheelwork to effect the same increase of power. Though we have only two moving parts in the first case to perhaps four or five in the second, yet, on investigation, we shall most likely find that the loss of power resulting from friction in the action of the first, will be considerably more than that involved in the action of the latter.

For reasons already stated elsewhere, we should make the motion of the parts of machinery as much as possible continuously rotatory, and their velocity as great as circumstances will permit of; as it is evident that to transmit the same mechanical effect with a greater velocity will require a proportionately less momentary pressure, and consequently a less strength of material in the moving parts.

CHAPTER VI.

OF THE SPECIAL CONSTRUCTION OF MECHANICAL COMBINATIONS.

71. Having considered thus far the different circumstances influencing the motion of machinery, as also the general principles involved in its construction, we now proceed to investigate in detail the different elements of mechanism which meet with general application in machinery, with a view to ascertain the principles upon which their action depends, as also the requisite strength and exact form of their component parts. In so doing, we shall confine ourselves to such constructions as are most commonly used, leaving the description of special constructions till we treat of the particular machines in which they occur.

72. *Shafting* constitutes the principal element of that part, in a system of machinery which we have termed the transmitter of power. It serves to convey the force which is generated in the engine to the different working machines, for which purpose it is provided with drums and belts, or else cog-wheels firmly keyed on.

The material of which shafting is generally constructed is, for ordinary sizes, wrought-iron; for large sizes, and where the inertia in the moving parts is rather an advantage than otherwise, as for instance in rolling mills, cast-iron is frequently employed. Shafts of wood were in common use before iron was manufactured, and applied with such facility as it is at the present day, but the former material is now almost completely abandoned ; the degree of accuracy now required of our machines, not being attainable in its employment, besides its being of a very cumbersome and ungainly appearance. The axles of wooden water-wheels are however still made of wood, the gudgeons being of iron and let into the same in a cross shape, strong wrought-iron hoops being driven on to keep them in firm.

As regards the arrangement of shafting and the proportioning of its diameter according to the amount of power to be transmitted by it, too much care and consideration cannot be bestowed on this subject. For where we have to deal with long lines of shafting, as is generally the case in mill work, it is a matter of very great importance, whether, in giving off the power to the different machines, we have, by oversight, placed those machines requiring most power furthest away from the prime mover, thus necessitating the diameter of the whole line of shafting to be made proportionately to the power required by these machines, whereas, had we placed them close to the prime mover, only a short length of shafting would have been of a large diameter, the rest of the length decreasing as the power is given off on to the lighter machines. The question here involved, is not merely the prime cost of the extra material, but the daily profitless expenditure of power in giving motion to useless mass. Again, if, through misdirected economy we make a shaft too weak for the power it has to transmit, vibrations and consequent loss of power and deterioration in the same will ensue in the bearings, and the gearing fixed upon the shaft.

The velocity with which the shafting should revolve .. also an important matter for consideration. As a general rule, with regard to economy of construction, we should let it go at a fast speed, as the diameter will, according to ..at we have stated elsewhere, decrease in inverse proporti.. the number of revolutions at which it is driven, the t.. mitted power remaining the same. At the same time w..

must take into consideration the velocity of the machines to be driven, and avoid driving the shafting at such a speed as would necessitate gearing of extravagant proportions or a complication of reduplications, involving loss of power, to arrive at the proper speed for the machine.

The form of shafting is generally solid, cylindrical, sometimes hollow, or in cases where it has to withstand great lateral pressure, it is provided with stiffening ribs or feathers as they are termed. Fig. 1, Plate III., shows the most general construction a length of shafting obtains. It has a boss at *a*, on which the gearing is keyed without weakening the strength of the shaft. The bearing or journal *b* should always be placed as near as possible to the gearing; it is turned very accurately and is of the same diameter as the body of the shaft, its length varying from one to one-and-a-half times its diameter. The coupling *c*, or part where the different lengths are connected together, should also have at least one bearing close to it, to prevent a sinking of the shaft.

73. With regard to the strength of shafting, it is obvious that it may be subjected to two strains, one tending to twist it, or *torsion*, the other a transverse strain caused by the weight or the leverage in the action of the gearing.

These two strains may either occur singly or acting together in a shaft, the latter being most generally the case. For either strains the strength of shafts varies as the cube of their diameters, that is, if a shaft of 2 inches diameter will transmit a certain power, or bear a certain cross strain, then another 6 inch diameter will transmit or bear a strain 9 times greater, or if $s, s' =$ the strain transmitted by two shafts whose diameters $= d, d'$ then

$$s : s' = d^3 : d'^3$$

from which

$$d = d' \sqrt[3]{\frac{s}{s'}} \cdot$$

With reference to torsion, the cube of the diameter of a shaft is directly as the horse power to be transmitted, and inversely as the number of revolutions per minute. Hence we find that for all similar proportions of horse power to number of revolutions, that is, whether it be 2 HP and 3 revolutions, or 20 HP and 30 revolutions, the diameter of the shaft will be the same.

To find the diameter of a wrought-iron shaft in inches when acting by torsion, and also with an ordinary amount of cross strain, such as may be caused by the leverage and weight of moderate sized gearing, the result of numerous experiments have proved, that if we multiply the quotient of HP ÷ revolutions by 300 we shall arrive at the correct value of d^3, or if d = diameter, HP = horse power, and r = number of revolutions per minute,

$$d = \sqrt[3]{\frac{\text{HP}}{r}} \times 300.$$

From this formula the following table has been computed, in which the ratios of HP to r are given, and the corresponding diameter of the shaft in inches and parts of inches:

$\frac{\text{HP}}{r}$	d	$\frac{\text{HP}}{r}$	d	$\frac{\text{HP}}{r}$	d
0·0266	2	0·5545	5½	2·430	9
0·0380	2¼	0·6337	5¾	2·638	9¼
0·0520	2½	0·7200	6	2·857	9½
0·0691	2¾	0·8138	6¼	3·089	9¾
0·0900	3	0·9154	6½	3·333	10
0·1111	3¼	1·	6¾	3·589	10¼
0·1429	3½	1·143	7	3·858	10½
0·1764	3¾	1·270	7¼	4·131	10¾
0·2133	4	1·406	7½	4·437	11
0·2558	4¼	1·551	7¾	4·746	11¼
0·3034	4½	1·707	8	5·069	11½
0·3575	4¾	1·871	8¼	5·407	11¾
0·4166	5	2·047	8½	5·760	12
0·4823	5¼	2·233	8¾		

Example: A shaft has to transmit 10 HP at the rate of 30 revolutions per minute: then 10 : 30 = ·333 and d = say 4⅝ inches. A shaft 6¾ inches in diameter can transmit 3 HP with

3 revolutions or 20 H P with 20 revolutions. Shafts not acting with torsion, but having to withstand considerable lateral strain, either from acting as a fulcrum or from carrying a great weight, as in fly-wheel shafts, must be calculated as a beam resting on both ends, or, if d = diameter in inches in centre of shaft, l = length between bearings in inches, and P = weight in cwts that it has to carry, reduced to the centre between bearings: then

$$\frac{d^3 \times 21}{l} = P$$

$$d = \sqrt[3]{\frac{l\,P}{21}}$$

giving a strength equal to six times the strain. This section may be decreased towards the bearings according to the form of the parabola.

74. With respect to the strength of the bearings, it is obvious that in shafting, transmitting strain by torsion, these should be as nearly as possible of the same diameter as the main body of the shaft. In shafting, subject to transverse strain, we shall arrive at the necessary diameter for the bearing in the following manner:

Let Fig. 2, Plate III., represent a journal of a shaft supported at both ends, and carrying a weight in the centre equal to 2P, then the journal is pressing on the bearing surface, and consequently the bearing surface is also pressing on the journal with a force equal to P distributed along the whole length. This force P we may consider as acting upwards in the centre of the journal, then the strength of the same may be calculated as that of a canter lever of the diameter d and length l, with a force P acting at $\frac{1}{2}\,l$, and we shall have

$$\overset{\text{cwt.}}{6 \times P} \times \overset{\text{ins.}}{\tfrac{1}{2}\,l} = \overset{\text{ins.}}{d^3} \times 31{\cdot}5.$$

$$d = \sqrt[3]{\frac{P\,l}{2{\cdot}6}}$$

for wrought iron, and

$$d = \sqrt[3]{\frac{P\,l}{2{\cdot}45}}$$

for cast iron, giving a strength equal to six times the strain.

For upright shafts, the diameter of the bottom journal, which has to carry the weight of the shaft and gearing, should be determined by the amount of pressure which may be allowed to come upon the square inch of bearing surface; this should never exceed 1000 lbs., and should, where practicable, be limited to half that amount. This limiting pressure must also be taken into consideration in determining the length and diameter of horizontal bearings, in which case we may allow as bearing surface one-third of the circumference.*

75. *Couplings.*—Although shafting is sometimes, where circumstances require it, forged together to considerable lengths, as in the case of shafting to travellers (Mr. Whitworth has in his workshop a shaft of this description, 3 inches diameter, 102 feet long, with a key bed groove cut in the whole length), yet generally for the sake of economy and facility in mounting, it is made in convenient lengths, and these are connected by means of what are termed couplings, consisting in general of a gland fitting accurately on the end of each length, and effecting a transmission of the strain from one into the other, either by their being half lapped over each other, as shown in Fig. 1, Plate II., or else, as in Fig. 2, by means of a key passing through the whole length of the coupling. The following are the dimensions according to Manchester rules, for various sizes of half lap couplings:

* For further investigations into the strength of shafting, see "Buchanan on Machinery," page 173; "Engineer and Machinist's Assistant," page 130.

Diameter of shaft.	Diameter of coupling.	Length of lap.	Length of box.	Diameter of box.
inches.	inches.	inches.	inches.	inches.
2	3¼	2¼	5½	5½
3	4¾	3½	7½	7¾
4	6	4	8½	9½
5	7¼	5	9½	11¼
6	9	6	12	13½
8	12	8	16½	18
10	14	10	18½	22
12	17½	12	21	26

For butt couplings the length of the box is generally made 4 times the diameter of the shaft, and the thickness equal to half the diameter. One object of couplings is the disengaging at any time of part of the shafting when the machinery driven by it is not required to be at work. This is sometimes effected by the driving back of the ordinary coupling boxes, but more generally where the disengaging occurs frequently, clutch couplings, as at Fig. 3, or friction cones, as shown at Fig. 4, are used. In these cases there must be a bearing on each side of the coupling; the disengaging is effected by means of a forked lever, one end of which fits round the groove *a*, the other, of sufficient length to afford leverage, forms the handle. The part *b* is keyed fast on the end of shaft *c*, while the other part *d* slides backwards and forwards on the end of shaft *e*, but is turned with the same by means of a solid feather *f* on the shaft.

The arrangement at Fig. 4 has the advantage over the clutch that it can be more readily engaged or disengaged without having to stop the driving shaft. If this is done with the clutch, the sudden shock in starting the machinery at the full speed of the driving shaft would probably cause some damage to be done. It is obvious that with the other arrangement the friction on the surfaces of the cones may be regulated so as to give a very gradual speed at starting.

These couplings are very frequently used in immediate connection with gearing, when, instead of the shafting not being continuous, the part of the coupling connected with the gearing turns loose on the shaft, while the other turns with the shaft, but is capable of sliding on the same as already described. Besides those mentioned, there are other varieties of couplings, among which the universal or Hooke's joint is particularly deserving of notice, though, on account of its costly construction, it is rarely applied excepting in cases where the two shafts stand at a slight angle to one another, which angle however must not exceed 15°, else the motion in the driven shaft becomes very irregular. Fig. 3, Pl. III., shows the construction of the same.

76. *Plummer Blocks, Brackets, &c.*—For the correct and economical working of machinery, it is essentially necessary that the different parts should be held in as nearly a true and rigid a manner as possible, in order that the geometrical proportions of the various moving parts should remain constant, and the loss of power attendant on their friction and vibrations be reduced to a minimum. What important items these latter are, even under the most favorable circumstances, will be readily understood when we mention that the number of horse power consumed by these and the inertia of the moving parts alone amounted to 75 in a well constructed saw-mill, the total power required to drive which, when all the work was on, was 250 H P, the friction therefore amounting to 30 per cent. on the whole.

It is therefore of the utmost importance that the supports for the bearings of shafting (plummer blocks) should fit the same with the utmost accuracy possible, and that the metal of which they are constructed should be of such a nature that, when properly lubricated, there shall be the smallest amount of friction between the surfaces in contact.

The form a plummer block generally obtains is shown at Figs. 5 and 6, Plate II., and an improved form at Figs. 7—9.

There is generally a bed plate as shown in the latter case, which is firmly fixed to the foundations. The bolt holes in the same and the corresponding ones in the plummer block by which the two are secured together, are slotted in reverse ways, so as to allow a certain amount of adjustment each way, for which purpose also, keys are inserted between the snugs on the bed plate.

The brasses, which form the part immediately in contact with the shaft, are made of a hard alloy of copper and tin, which has been found to give the smallest amount of friction with iron. In Fig. 6 they are made of an octagonal shape to prevent turning round, and are provided with flanges. For the sake of economy they are made of a greater thickness at top and bottom, where the greatest amount of wear generally takes place, and there is a space between the two halves, as well as between the cover and the body of the plummer-block, to allow of an adjustment of the brasses as they wear out. An arrangement is made in the cover for the proper lubrication of the bearing surfaces.

The form of plummer-block as shown in Fig. 7 has the advantage over the ordinary form, that it does not necessitate such a great degree of accuracy in fixing, inasmuch as the outer surfaces of the brasses are spherical, thus allowing a certain amount of motion every way, without impairing the rigidity of the construction. It is particularly adapted for long lengths of shafting going at a quick speed, where the maintaining of a perfectly true geometrical axis is next to impossible, and where any inaccuracy in the fitting of the brasses to the bearing would cause a great amount of friction and wear them out rapidly. The introduction of this form of brasses is due, we believe, to Mr. Bodmer. Spherical bearings, with which these must not be confounded (as the bearing surfaces are still cylindrical), have proved to be of no practical value in mill-gearing, on account of the unequal distribution of pressure on the bearing surfaces, and consequent rapid wear of the same.

Bearings that have only a partial rotatory or an oscillating motion, as in those to the beam of a steam engine, should be occasionally disconnected, and turned round a certain amount, else they are apt to wear uncylindrical.

Bearings to upright shafts or cup-bearings are still more difficult to construct, so as to fit accurately, than those to horizontal shafts. For in the former case, we have not only to insure the axis of the shafts coinciding with that of the cylindrical part of the cup, but also that the bottom or principal bearing surfaces fit one another truly. For all such cases, therefore, where the shaft moves with a great velocity and experiences considerable pressure, the spherical principle should be adopted, as shown in Fig. 10, Plate II., which represents a bearing particularly adapted for turbine shafts.

The bed-plate *a*, which is bolted to the foundations, has a spherical socket into which fits the spherical end of the cup *b*; this allows a self-adjustment of the cup to suit the direction of the shaft. A further adjustment is obtained by making the part *e* spherical, so that the bearing surfaces are always in accurate contact. The parts *e* and *f* are of brass, and *d* is of cast-steel, and is screwed into the shaft *c*. The pipe *g* admits the oil for lubrication, which passes down grooves in the cylindrical bearing, and out at *h*. The perfect lubrication of these bearings is a most important point; as an example, we may state that we have witnessed several cases where, on account of this and the bearing surfaces having been defective, the two bottom bearing surfaces have been found as firmly welded together, though under water, as they could have been in a smith's fire.

77. *Brackets.*—These are castings adapted to the bearings of shafts, either in connection with, or instead of ordinary plummer-blocks. They are termed according to the mode of their application:—

1. Pendant brackets, when the shafting is suspended from the ceiling or elsewhere. Figs. 11—13, Plate II., indicate a pendant bracket for three shafts meeting together.

2. Wall brackets, when fixed against a perpendicular wall as shown at Figs. 14—16.

3. Wall boxes, when the shaft passes through a wall, a casting resembling four sides of a box being inserted in the wall, forming the bed-plate for the plummer block.

4. Pedestal brackets, which are generally adapted where we require a bearing for a vertical shaft, immediately below which runs a horizontal shaft, as shown at Fig. 17.

Besides the above-mentioned methods for supporting shafting, we may mention one which, though at present confined in its application to wheels moving round fixed axes, deserves, from its anti-friction qualities, to be also adapted to the bearings for shafts, particularly where these are of a large diameter, and do not run at too great a speed. It consists of supporting the shaft by means of a series of small rollers as shown in Fig. 4, Plate III., these being turned accurately to agree with the difference of the radii of the shaft and the surrounding box. The friction here is entirely reduced to that of rolling contact.

78. *Bolts and Nuts.*—These represent in general the medium through which the engineer brings about firm connection of different parts of machinery. In their general form, they obtain a solid *head* forged on to one end of the bolt, while the other has a *thread* cut on a certain portion of its length, on to which is screwed a *nut*. The form, however, is varied, according to the manner of application. Figs. 18—26, Plate II., represent the most usual forms of bolts. Fig. 18 is termed a foundation bolt, being generally of considerable length, so as to pass through and hold on to a certain mass of masonry, the weight and inertia of which has to resist the strains in the framing, &c., which is held down by these bolts. Holes are left in the masonry down which they are passed, and at the bottom of the same are built what are termed pigeon holes, accessible to the workman, who there places on the *washer* and inserts the *cotter* into the end of the bolt.

Fig. 19 is termed a T *headed* bolt, and is very useful in cases where we require to fasten against a plate which is only accessible from one side. In this case a slotted hole is made in the plate, of the size of the T head of the bolt, which is then passed through and turned round at right angles to the hole.

Fig. 20 represents a *double-ended* bolt, for holding together three objects independently of each other. Fig. 21 is an eye-headed bolt, and is generally used for fixing the glands to stuffing boxes. Fig. 22 is termed a *lewis bolt*, used for fixing on to a block of stone; for this purpose a hole is cut in the stone, large enough at the top to admit the thick end of the wedge-shaped bolt, which is barbed at the angles and run with lead. Fig. 23 is the ordinary bolt and nut, being square under head to prevent the bolt turning round when being screwed. At Fig. 24 a thread is substituted for the head, and the bolt is *tapped* into the metal on to which it holds, for which purpose the part not threaded is flat on two sides. Fig. 25 is a *countersunk-headed*, and Fig. 26 a *round-headed bolt*.

With regard to the strength of bolts and nuts, in most cases the actual strain that comes upon them cannot be accurately determined; for wherever it is required to tighten up the nuts considerably, the strain put upon the bolt in so doing frequently by far exceeds the pull they are meant to withstand. For this reason it is the practice to allow a large margin for bolts of small diameter, calculating their sectional area at 2—3 tons per square inch; bolts of large diameter may be taken at 5—6 tons per square inch. Of course, in calculating the effective area, the diameter, less the depth of thread, must be taken. Where the bolts are long, it is the practice to swell the part on which the thread has to be cut, to obviate a waste of metal. The depth and diameter of nut, which is hexagon-shaped, differs with different engineers. In general, the depth rather exceeds the diameter of the bolt, and the larger diameter of the nut is rather less than twice that diameter. In cases where there is not a steady pull upon the bolt, but where it has to withstand jerks, *lock nuts* should always be used, as shown in Fig. 7, the larger being at the top, and not the reverse, as generally takes place; because it is very evident that if there is any inaccuracy in the fitting of the threads, the top nut has to bear the whole of the strain.

79. *Contrivances for modifying Motion.*—Under the term "modifying motion" we imply a change from one description of motion to another. The descriptions of motion we generally find in machinery are, rectilinear, continuous and alternate; and circular, continuous and alternate. The changes that can take place between these kinds of motion may be expressed as follows, where ⟶ means rectilinear; ⟷ rectilinear alternate; ↻ circular continuous; ↶↷ circular alternate:

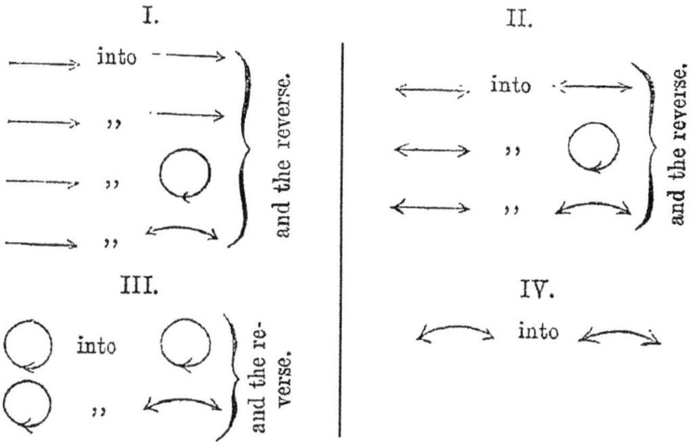

Many of these combinations are, however, of little practical value, by far the greater number of mechanical contrivances originating in two only of them, namely, ↻ into ↻, and ↻ into ⟷; the modifications consisting chiefly in alterations of the velocity-ratios of the parts in connection.

80. *Of the Crank.*—The crank and connecting rod form the most common means for converting continuous circular into alternate rectilinear motion. The end *b* (Fig. 5, Plate III.) of the connecting rod is caused to slide backwards and forwards in a straight line, while the other end, *a*, connected with the crank pin, revolves with the same in the circle *a*, 1, 2, 3, &c., the length of whose diameter evidently expresses the *length of stroke* of the other end of the connecting rod, *b*. The motion produced is of such a nature, that if the velocity of either point, *a* or *b*, be constant, the velocity of the other point, *b* or *a*, must be variable. Were the connecting rod of infinite length, then, if the point *b* pass through equal spaces in equal times, the projections of the corresponding spaces passed through by *a* on to the diameter, must be equal, as in Fig. 6; and if the spaces passed through by *a* on the circumference of the circle are equal, then those passed through by *b* must correspond with the projections of the former on the diameter. If $r =$ radius of the crank, l the length of connecting rod, and a the angle at which the crank stands, then to find X, the distance between *c* and *b*, we have:

$$X = r \times \cos. a + l \quad . \quad . \quad . \quad . \qquad \text{I.}$$

But in practice, the connecting rod is generally of such a length as to cause a still greater irregularity in the relative velocities of the points *a* and *b*, inasmuch as, instead of the distance between *a* and *b* being constant, as we have before supposed it to be (in taking this distance as infinitely long, and thereby considering all the positions of the connecting rod parallel to one another), it now varies to the extent of the *cosine* of whatever angle *a b* may form with the horizontal line. It is therefore evident that to reduce this increase of irregularity in the motion to a minimum, we should strive to make the connecting rod as long as possible.

If β represent the angle formed by the connecting rod at any part of the stroke, and the other significations are as above, we have for the distance at any time between the points c and b:

$$X = r \times \cos. \alpha + l \times \cos. \beta . \qquad \text{II.}$$

or, as

$$r \times \text{sine } \alpha = l \times \text{sine } \beta, \text{—}$$

$$\text{sine } \beta = \frac{r}{l} \times \text{sine } \alpha .$$

$$\cos. \beta = \sqrt{1 - \left(\frac{r}{l}\right)^2 \times \text{sine}^2 \alpha}$$

and

$$X = r \times \cos. \alpha + l \sqrt{1 - \left(\frac{r}{l}\right)^2 \times \text{sine}^2 \alpha} . \qquad \text{III.}$$

Fig. 7, represents the motion of the end b of the connecting rod, the horizontal divisions being equal parts of one revolution of the crank, and the ordinates representing the corresponding distances passed through by the end of the connecting rod.

The gradual increase and decrease of velocity in the connecting rod at the beginning and end of the stroke, due to the uniform velocity of the crank pin, is of great dynamical advantage in many cases; as for example, in pumps worked by cranks, where the body of water in the pipes is started and stopped by this means with an increasing and decreasing velocity, thus preventing the loss of power attendant on giving the whole mass of water suddenly a great velocity at the beginning, and stopping the same as suddenly at the end of the stroke, as in the Cornish pumping engines.

It is erroneously supposed that a great loss of power takes place in the action of the crank and connecting rod, owing to their oblique position to the direction of the power to be transmitted. A little investigation into the matter will show that the only loss that takes place is an absorption of a small amount of power by friction at certain points of the stroke, in the guides to the cross head, owing to a certain upward or downward pressure arising from the resolution, into two forces, of the motive power.

81. Fig. 1, Plate IV., shows the general construction and proportions a wrought-iron crank obtains. The strength, both of the body of the crank and the crank pin, has to be calculated according to the formula for canter levers; namely, if l is the length of the crank from outside of boss to centre of crank pin, P the breaking strain in pounds, b the breadth on the face, and t the thickness at the boss, then,

$$\frac{b^2 \times t \times 6000}{l} = P;$$

and for the crank pin, if l is the length from face of crank, P the breaking strain distributed over the whole length, and d the diameter of the pin, then,

$$\frac{d^3 \times 6000}{\frac{l}{2}} \times \cdot 589 = P.$$

82. *The excentric* is a form of the crank in which the crank pin obtains a greater diameter than the shaft. In this case the body of the crank is done away with, and the pin, in form of a sheave, either in one piece or in halves, is keyed directly on to the shaft. Round the circumference, between projecting flanges, is fitted the *excentric strap* of gun-metal, also in halves, furnished with ears, through which pass the ends of the forked excentric rod, as in Fig. 2, Plate IV. Means for lubricating the rubbing surfaces of strap and sheave must be provided, else the friction would soon cause their destruction. This arrangement is only adopted where the power is transmitted *from* the crank-shaft, and that only of a limited amount, as, for instance, for slide valves.

Wherever the power to be transmitted is considerable, the shaft will either be cranked, or a boss forged upon it to which is given the necessary excentricity. Fig. 8, Plate III., shows a construction by means of which the *throw* of the excentric may be altered at pleasure. A variable velocity may be given to the crank by a combination of the crank and the excentric as shown at Fig. 3, Plate IV., which ingenious arrangement was adopted in Mr. Whitworth's workshops for giving to the table of the planing machines a quick motion for the return stroke. The boss *a*, on which turns the spur wheel *b*, is fixed. The crank *c* turns with the spindle *d*, which is fitted excentrically in the boss *a*. The pin *e* with the block *f* are fixed to the spur wheel, which is set in motion by the pinion *g*. In revolving round, this wheel carries with it the crank, but owing to the excentricity of the two centres, the block *f* slides down the slot in the crank, and in so doing approaches nearer to the centre on which the latter revolves. This has the effect that, as the angular velocity of the spur wheel is constant, it will cause different points in the radius of the crank, *consecutively* to revolve with the same linear velocity; or, in other words, will cause the angular velocity of the crank gradually to *increase* during one half of the revolution, and gradually to *decrease* during the other half. The stroke of the crank may also be varied by shifting the crank pin in the slot *h*.

83. *Cams.*—This term is applied to revolving discs fixed on shafts, the form of the periphery of which is such as to impart to a lever or rod in contact with it, any velocity or description of motion that may be required. It is by means of these ingenious contrivances that we are enabled in machines to bring about those automaton motions which act as a substitute for that inimitable part of nature's machinery, the arm of man. Fig. 4, Plate IV., represents a form of cam which gives that which we found wanting in the motion of the crank and excentric, namely, the conversion of uniform rotatory motion into uniform rectilinear reciprocating motion. The principle of its construction is, that for each successive equal part of one half revolution of the cam, the rod must have been moved by its periphery

through a corresponding equal part of the entire stroke. If *a e* represent the stroke required, then divide the same into any number of equal parts *a—b*, *b—c*, *c—d*, *d—e*, and describe from the centre of rotation circles passing through these points of division; divide half the circumference of these into as many corresponding equal parts, and draw radial lines through the points of division *e f g h i*, then a^1 a^2 a^3 a^4 will represent the successive positions of the point of the rod as the cam revolves, and of course for the return stroke these positions will follow in the reverse order. To lessen the friction the rod is provided with a roller, to suit which two equidistant curves are made, forming a groove in which the roller runs. On account of the form it obtains, this is generally called the heart-shaped cam.

Fig. 9, Plate III., represents the motion of the rod to this cam; as before, the horizontal distances represent equal parts of one revolution, while the ordinates are the corresponding distances passed through by the roller.

Fig. 10, Plate III., represents a form of cam, the action of which resembles that of the crank or excentric; it is formed so as to impart to the rod a uniform increasing velocity during one half the stroke, and a uniform decreasing velocity during the other half. Here the parts 1—3, 3—5, &c., instead of being equal, are in the same proportion to each other as the distances fallen through in the first, second, third, &c. second of time, by a body acted upon by gravity. These will, therefore, if we consider the distance passed through during the first part as a unit, be represented by the figures 0, 1, 3, 5, 7, 9, 11, 13, &c. If we draw circles through these divisions of the stroke, and divide one-quarter of the circumference into a corresponding number of equal parts, draw the radial lines, then the points a^1 a^2 a^3 will represent the respective positions of the centre of the roller at equal parts of the revolution, and an equidistant curve to suit the size of the roller will give the form of the cam.

Fig. 11, Plate III., represents the motion of the roller

during one half revolution of this cam; the abscissa and ordinates are as before described.

This form of cam, but modified as in Fig. 12, is frequently used for lifting the expansion valves in steam engines. Here a revolution of the cam gives two lifts instead of one, and the rod or lever remains stationary for a certain period at the end of each up and down stroke, this time being regulated according as it is required to cut off the steam at $\frac{1}{2}$, $\frac{1}{3}$, $\frac{1}{4}$, &c. of the stroke of the piston.

Another form of cam, much used in working the slide valves of steam engines, is shown in Fig. 13. The kind of motion obtained here, is, as will be seen from the diagram, Fig. 14, a quick up and down stroke, and a rest at the end of each stroke, equal to one-sixth of a revolution.

As will be readily understood, innumerable varieties of motion may be obtained by different forms of cams. We shall have to refer again to the subject in the description of machines where they are introduced for some specific purpose, and shall now only mention two general rules to be observed in the setting out of cams, namely:—

1. That the form of the cam must be such that wherever the same is exercising a pressure against the rod or lever which it has to move, the direction of this pressure must be as nearly as practicable in the direction of the motion which it has to impart; or in other words, the surface of the cam must be as nearly as possible perpendicular to the line of motion, otherwise a considerable amount of power is lost in friction. This, however, does not apply to those parts of the cam where the same is retreating from the roller, which is kept up to it either by gravity or by the action of a spring; to these cases we must apply the second rule, which is, that wherever the surface of the cam drops suddenly, its form must be such that the roller is always in contact with it, or, in other words: The surface must not retreat faster than the roller can follow it up, else the latter will come down with a blow upon the

former, which would act injuriously, not only upon the cam and lever, but also upon the working of the machine. There are, however, cams which receive the name of wipers or tappets, whose object is the raising up to a certain height and then letting fall of hammers or stampers, but these in falling do not come in contact with the cam. The form that such cams generally obtain is that of the involute, any part of which curve is at right angles to its corresponding tangent from the generating circle, and consequently very nearly at right angles to the direction of motion.

84. *Of Drums and Belts.*—These represent the most general method of transferring power by rotatory motion from one shaft to another. They are employed now much more frequently than formerly, wherever practicable, in preference to cog-wheels, on account of several advantages which they possess over the latter. In the first place the driving and driven shafts may be at almost any distance, within reasonable limits, from one another, which in cog-wheel gearing would entail a great expenditure and loss of power through the introduction of intermediate wheels. 2°. By means of the contrivance known as the fast and loose pulley, the engaging and disengaging of machinery takes place much easier and with less liability to shocks, owing to the inertia of the moving parts, than is the case where cog-wheels are used, the arrangement being besides much simpler, and inexpensive. 3°. While at work, the motion is much easier, owing to the absence of the vibrations that will occur, even in the best constructed wheel-work, from the inexact working of the teeth.

The principle from which drums and belts derive their power is that of friction. A continuous leathern belt is wrapped round the two drums or pulleys, being generally in contact with about one-half the circumference of each. The adhesion of the belt to the pulley, or the friction between the two surfaces, regulates the amount of power that may be transmitted, and this friction is again dependent on the force with which the belt is strained over the drums, and also on the amount of surface in contact. If P represents the force at the circumference of the driving drum to be transmitted, f the force with which the belt is strained to produce the proper amount of friction, c the coefficient of friction, s the length of belt in contact, r the radius of the drum, $e = 2{\cdot}71828$ the basis of the hyperbolic logarithms, then we have:

$$P = f \times e^{\frac{cs}{r}} \quad . \quad . \quad . \quad . \quad . \quad \text{I.}$$

or :

$$\log P = \log f \times \frac{cs}{r} \times \log e \quad . \quad . \quad . \quad \text{II.}$$

and if we substitute for c ·30 as the coefficient of friction for leather on iron, and as log e (basis 10) = ·434, then:

$$\log P = \log. f + {\cdot}13 \times \frac{s}{r} \quad . \quad . \quad . \quad \text{III.}$$

85. The following table gives, for different proportions of s to the circumference of the drum, the corresponding values of $e^{\frac{cs}{r}}$ in which $c = {\cdot}28$ for greased leather on iron, and $c = {\cdot}47$ for the same on wood.

$\frac{s}{\text{circumf.}}$	Values of $e^{\frac{cs}{r}}$	
	Leather on wood.	Leather on iron.
·2	1·80	1·42
·3	2·43	1·69
·4	3·26	2·02
·5	4·81	2·41
·6	6·59	2·87
·7	9·00	3·43
·8	12·34	4·09
·9	16·90	4·87
1·0	23·14	5·81

Generally s is about half the circumference. Where one drum is smaller than the other, it is rather less than this on the small one, and rather more on the larger. Where the belt is crossed it is more on either.

As regards the width of the belt, this will be found ample with respect to friction if we calculate the cross section of the same for the strain to be transmitted, in which case one-eighth of an inch square is allowed for every 5 lbs. strain. The drum is made about one-fourth wider than the belt and is slightly rounded on the circumference, so as to give the belt rather a greater amount of tension in the centre, which prevents the same from running off the drum, owing to its propensity to run *on to* instead of *off* a part where it is strained more than at the other.

To arrive at the strain in the belt due to a certain power required to be transmitted, we must divide the HP or lbs. feet by the surface speed of the drum in feet per minute, which will give the strain in pounds. For instance, a drum four feet diameter is required to transmit 3 HP with 60 revolutions per minute:

Circumference of 4 ft. diameter =	12·5 ft.
Revolutions per minute =	60
	750 ft.
33,000 lbs. ft. × 3 =	99000 lbs. ft.
99,000 lbs. ft. : 750 =	132 lbs. strain.

To this must be added the force with which the belt is strained to produce the necessary amount of friction. If the length of belt in contact is equal to half the circumference, we have according to the table:

$$e^{\frac{c\,s}{r}} = 2{\cdot}41,$$

and from equation I.

$$f = \frac{P}{2{\cdot}41}$$

$$= \frac{132}{2{\cdot}41} = 54\tfrac{1}{2} \text{ lbs.}$$

and 132 + 54 = 186 lbs. total strain in the *leading* belt,

the *following* belt being only strained with 54 lbs. The section of belt due to the 186 lbs. will be 186 lbs. : 5 lbs. = 35 square eighths of an inch, and if the belt is $\frac{1}{8}$ inch thick, the width of the same will be 35 : 8 = $4\frac{3}{8}$ inches.

86. In transmitting power from one shaft to another, it is most generally required to change the speed. This is effected by making the diameters of the drums inversely proportioned as the speeds of the two shafts; or, the surface speed of the two drums being obviously the same, the products of their circumferences with their respective numbers of revolutions must be equal. For instance, a shaft moving at the rate of 80 revolutions per minute, with a drum 3 feet diameter, has to drive another shaft at the rate of 120 revolutions per minute; what will be the size of the drum for this purpose?

$$3 \text{ ft.} \times 80 \text{ revs.} = x \text{ ft.} \times 120 \text{ revs.}$$

$$x = \frac{240}{120} = 2 \text{ ft. diameter.}$$

Figs. 5—10, Plate IV., show various methods of driving drums. Fig. 5 represents the direct method, where the drums are in the same plane and the axes parallel. In Fig. 6 the belt is *crossed*, which has the effect of making the drums revolve in contrary directions. The belt is twisted in such a manner, that if one of the drums were turned round in a direction vertical to the axis, it would be in the same position as at Fig. 5. Fig. 7 is a case where the axes are parallel, but the drums not in the same plane: here it is necessary to lead the belt over guide pulleys. Fig. 8 is a similar case, where the axes are in the same plane but not parallel. In Fig. 9 the axes are neither in the same plane nor parallel, yet in this case no guide pulleys are needed. Fig. 10 represents a similar case where guide pulleys are required.

It sometimes occurs that, besides the rotatory motion, the driven drum has imparted to it an independent motion, either along its shaft, or together with the shaft. In the first case the driving drum must be as wide as the whole lateral motion of the driven one; in the second case the drum can only move in the circumference of a circle whose radius is equal to the distance of centres of the two drums. By means of the arrangement as shown in Fig. 15, the distance of centres may also vary.

87. *Arrangements for varying the velocity-ratio of Shafts by means of Drums.*—This may require to take place: 1, occasionally, that is, not while the machinery is in motion; 2, gradually, while the machinery is in motion; and 3, suddenly, while the machinery is in motion.

The first is most generally effected by means of *speed pulleys*, consisting of a series of drums of different diameters, generally formed in one casting, as in Fig. 13, Plate IV., fixed in one order of succession on the driving shaft, and in the reverse order on the driven; the proportion of the respective diameters of each pair being such that the sum of the two is always constant, so that the same length of belt will do for all the speeds. Another arrangement by which an occasional change of speed may be effected, is represented in Fig. 11. Here the diameter of one of the drums is made variable, and the belt is kept strained by means of the weighted roller *a*. The part of the *expansion-drum* marked *b*, consisting of a boss and grooved arms, is keyed fast on the shaft; on to another portion of the arm *c*, which slides up and down, in the groove of *b*, is cast a portion of the circumference of the drum; it has also a stud *d*, fitting into the curved slot of the disc *e*, which moves loose on the boss *b*, and has teeth on its circumference into which works a pinion *f*, with ratchet fixed to the part *b*, and turned with a handle. As the disc is turned to the right or left, the studs move up or down in the curved slots, and the diameter of the drum is increased or diminished.

88. A gradual change of speed whilst the machinery is in motion is generally effected by means of two conical drums

placed in a reverse sense to one another, so that the sum of any two corresponding diameters is always a constant quantity. The belt is made by certain contrivances to move laterally along the surfaces of the cones, and thus, a gradual increase or decrease of speed in the driven drum is brought about, on account of the constantly varying proportion of the two diameters, while the angular velocity of the driving drum remains constant.

Fig. 12, Plate IV., shows an application of this contrivance to throstle spinning frames, for increasing the velocity of the same, as patented by Lister: *a* is a drum to which the motion is communicated, this is fast on the shaft *b*, to which is also fixed the conical drum *c*; *d* is the shaft communicating motion to the spinning frames, upon which is fixed, by means of the clutch *e*, the conical drum *f*, receiving its motion by means of the crossed strap *g* from *c*. The contrivance by means of which the strap *g* is caused to move along the surface of the drum, is as follows: *h* is a small pinion fixed on the shaft *b*, this turns the spur wheel *i*, which by means of the stud *k* gives motion to the lever *l* and ratchet-wheel *m*. This is fixed on the end of the screw *n*, and by the action of the same the screw is caused to revolve a certain portion of a revolution to every revolution of the spur-wheel *i*, and in so doing carries along the guide *o*, by which the strap *g* is moved. When the strap has arrived at the end of the drums, and it is required to bring it back again rapidly, the ratchets are thrown out of gear, and the drum *f* is enabled to revolve loose on *d*, by drawing the clutch *e* out of contact by means of the lever *p*; a quick reversing motion is then imparted to the screw by turning the spur-wheel *q* by hand. The whole machine may be thrown out of gear by leading the driving-belt on to the pulley *r*, revolving loose on the shaft *b*.

The alteration of speed by means of the conical drums may be made not to take place uniformly, by altering the shape of the cones as at Fig. 16, Plate III.; they will admit of any variation as long as the sum of the two corresponding diameters always remains the same, and the slope of the surface does not become too much for the strap. We have before mentioned that the strap has a tendency to run on to the part where it is strained most, so that, within certain limits, it will never slip off the slope of the drum.

89. A sudden change of speed is sometimes required; as, for instance, to obtain a quick running back motion. There are various ways in which this may be effected, of which Fig. 17, Plate III., is perhaps the most simple: *a* is the driving shaft, *b* is driven; on *a* are fixed the two drums *c* and *d*, of different sizes; on *b* are the fast and loose drums *e e′* and *f f′*, also of different sizes. The strap on *c* and *e* is direct, that on *d* and *f* is crossed, so as to reverse the motion. They are so arranged, that while the strap *c* is on *e*, the other, *d*, is on *f′*, that is, the loose drum; and again, when *d* and *f* are in communication *c* and *e* are out of gear. It is evident that the belts cannot be moved together, but that the one in gear must first be moved on to the loose drum, and then the other be moved into gear.

Fig. 14, Plate IV., represents another contrivance for giving what is termed *double speed* to the spindles of the self-acting mule. On the spindle *a*, which receives the motion, are three drums, *b*, *c*, and *d*; of these, *b* is fast on the spindle, *c* is loose, and *d* also loose, but connected with the pinion *e*, which works into the spur-wheel *f*, fixed on the spindle *g*, to which the motion is to be communicated. On the spindle *a* is fixed the spur-wheel *h*, which works into the pinion *i*, likewise fixed on the spindle *g*. When the belt from the driving drum *k* is on *b*, it communicates a fast motion to the spindle *g*; when on *c*, the machine is out of gear, and when on *d* it imparts a slow motion to *g*.

In changing a belt from a fast to a loose pulley, it is the *following*, not the *leading*, part of the belt which must be guided by the fork situated near to the fast and loose pulley.

It is also evident that the driving drum must be as wide as the fast and loose pulley together. These are generally made about one-fourth wider than their belts.

90. *Cog Wheels*, in the widest sense, may be described as: *Cylinders, or cones, revolving upon fixed axes, whose circumference is furnished with projecting parts, called teeth, of such a form, that in coming in contact with corresponding teeth on similar cylinders, or cones, on other fixed axes, they shall impart to these a rotatory motion with constant angular velocity, precisely the same as though the periphery of the cylinders or cones themselves were in contact*

According as the axes of the wheels are parallel, or at an angle to one another, these are termed *spur-wheels, bevil-wheels;* and when the axes are at an angle, but do not intersect one another, they are termed *hyperbolical wheels* or "*skew bevils.*"

The *pitch line*, or *pitch circle*, of a wheel, when considered relatively to the pitch circle of another wheel in contact with it, indicates the velocity-ratio of the two wheels; that is, the relative velocity of the two wheels is the same as that of two cylinders of the same diameters as the pitch circles, when rolling upon one another by friction. Upon this circle is set out the *pitch* of the teeth, or the distance from centre to centre of teeth. As in two wheels that work together, the distance from the face of one tooth to the corresponding face of another must in both wheels be alike, it is evident that *all wheels working together must have the same pitch.*

The *line of centres* is a straight line, passing through the centres of the axes of two wheels in contact. The part of the tooth above the pitch line is called the *face*, the part below is the same.

If Fig. 1, Plate V., represent two cog-wheels with straight faces, it will be easily seen that, supposing the wheel A to be the driver, and B the follower, then the velocity imparted to the wheel B by the tooth at *a* will not correspond with that imparted by the tooth at *c*: that in fact the angular velocity of A being constant, that imparted to B will gradually decrease from *a* till the surfaces in contact are exactly in the line of centres, will then suddenly increase to a certain extent, owing to the point of contact changing suddenly from *d* to *e*, and will then gradually decrease from *b* to *c*. Besides the inequality of the motion, this construction would have two other serious objections.

First, that, owing to this varying angular velocity, only one pair of cogs would be in contact at the same time, and when they left each other, there would be a sudden *slip* of the driven wheel till another pair came in contact.

Secondly, that there would be a very considerable amount of friction due to the edge of one cog sliding along the surface of the other.

In order, therefore, that cog-wheels should work accurately and economically, it will be necessary that the form of the teeth should be such as to fulfil the following conditions:—

1. The relative angular velocity of the two wheels must be constant.

2. There must be no jerk when two teeth come out of contact; that is, another pair of teeth must be already in contact.

3. The form of the teeth must be such that the surfaces in contact do not slide but roll upon one another.

The first-named condition will be complied with if we make the form of a tooth such that, supposing the circumferences of two ideal circles, which represent the relative proportion of two cog-wheels (pitch circles), to roll upon each other, it shall coincide with the curve described by a point in one pitch circle while rolling upon the other; or in other words, that it shall be part of the epicycloidal curves of which the pitch circles are the generating circles.

91. *The wheel and trundel.*—Let Fig. 2, Plate V., represent a pair of wheels, of which A is a toothed wheel but B

a trundle; that is, a wheel in which the teeth are replaced by cylindrical staves; and let us first suppose the diameter of these staves to be infinitely small or simply a point. Then the two pitch circles rolling upon each other, the distance C D on circle A will be equal to C E on circle B, or by the time a point on circle A at C has travelled to D, a corresponding point on circle B will have travelled to E, and the curve D E will represent the path of the latter during this portion of a revolution, being part of the epicycloidal curve, of which B is the generating circle. If, therefore, we form a tooth on circle A according to this curve, it will, while urging round circle B, by causing the pin E to slide along its surface, impart to the same a constant angular velocity, just the same as though the two pitch circles were rolling upon one another. As, however, the staves on the trundle B are in reality not mere points, we obtain a practical solution by taking the points on the pitch circle B, Fig. 33, as the centres of the circles representing the size of the staves, and then with the same radius forming an equidistant curve to D E by striking in the different positions of the stave while its centre is moving along the curve D E, as shown in Fig. 3, Plate V. To allow the centre of the staves to come down to the pitch circle A at the line of centres, the teeth on the latter must be hollowed out below the pitch line.

Owing to the friction caused by the staves rubbing along the faces of the teeth, abrasion ensues, rendering the action of this species of gearing inaccurate after a time. For this reason this construction is not much used at the present time.

92. *Teeth of wheels with epicycloidal faces and radial flanks.*—In the place of the cylindrical staves just mentioned, we will now furnish the wheel B, Fig. 4, with teeth, whose flanks, or the portions of the teeth below pitch line, are radial lines. We have now to find the proper form for the faces of the teeth of the wheel A, so that the velocity-ratio of the two wheels shall be constant.

On the line of centres, the point of contact between the radial line and the curve C F, must evidently be on the pitch line, but as the radial flank moves into the position E G, the curve D G must, for the action to take place with as little friction as possible, have rolled up the line E G, and the problem we have now to solve is to find the radius of the generating circle for the epicycloidal curve D G, being furnished with the following data:—1. For the velocity ratio to be constant, the portion of the pitch circle A = *d c* must be equal to the length E C on B. 2. The length G C on the generating circle must be equal to *d c* on A, and therefore; 3. The length G C on the generating circle must also be equal to E C on pitch circle B.

Now this last condition, in other words, says, that the generating circle rolling *inside* the pitch circle B, must form, as portion of its hypocycloidal curve, the radial line E G, and we know that there is only one circle which, rolling inside another, will form for its hypocycloid a line corresponding with the diameter of the base circle, and that is a generating circle whose diameter is equal to half that of the base circle. The diameter of the generating circle for the faces of teeth of the wheel A, is therefore half that of the pitch circle B. This can also be proved in the following manner:—

$$C D = C E = C G$$
$$\therefore R \times \angle \alpha = r \times \angle \beta,$$

which condition can only be fulfilled when

$$\angle \alpha = \tfrac{1}{2} \angle \beta, \text{ and } r = \tfrac{1}{2} R.$$

In the same manner we shall find that the generating circle for the curve of the faces of the teeth for B, which come into contact with the flanks of the teeth for A before the line of centres, will be also half the diameter of the pitch circle A.

93. *Teeth of wheels with epicycloidal faces and hypo-*

cycloidal flanks.—We have, in the foregoing, noticed a special case where the hypocycloid of the generating circle for the faces of A corresponded with a radial line of B. To obtain a general solution, we now state that *the corresponding flanks and faces of two teeth in contact will always work correctly when they have both the same generating circle, forming respectively the hypocycloid in one pitch circle for the flank, and the epicycloid on the other pitch circle for the face, the diameter of this generating ci. le being either equal or less than the radius of the pitch circle of which it forms the hypocycloid.*

In Fig. 5, R′, and r', are optional, but less than $\frac{1}{2}$ R and $\frac{1}{2}$ r, D G = epicycloid of r', E G = hypocycloid of r', H I = epicycloid of R′, K I = hypocycloid of R′.

This is the form commonly given to the flanks and faces of teeth, though the true curves are generally replaced by aproximative portions of a circle. For a description of the method for finding the radii of these circles we must refer the reader to Professor Willis's "Principles of Mechanism," 2nd Edition, where will also be found a method for constructing a corresponding flank to any form of face, and *vice versâ.*

94. *Teeth of wheels formed with involute curves.*—If in Fig. 6, A and B represent the pitch circles of two wheels in contact, then if we describe from the centre of A a circle A′ of optional diameter, but smaller than that of the pitch circle; if we then draw a tangent to this circle passing through the point of contact C in the line of centres O O′ and draw a perpendicular E O′ to the tangent D C E, so that it shall pass through the centre O′ of pitch circle B; this perpendicular will represent the radius of a circle B′ which shall be of the same proportion to A′ as B is to A, and the line D C E will be a tangent to both these circles. Let us now consider this tangent a flexible line, and from the point C describe first the involute of A and then the involute of B; these two curves will then represent

the faces of two teeth that will act correctly upon one another, for:

(1.) By virtue of the principle of the involute, if the wheels make a portion of a revolution, then the point of contact C will have travelled along the tangent D E to C′, and the tangent will, in all positions, be a normal to the point of contact passing through the line of centres at C.

(2.)
$$C\,C' = F\,F' = H\,H'$$
$$G\,G' : I\,I' = F\,F' : H\,H'$$
$$\therefore G\,G' = I\,I'$$

i. e., the velocity-ratio is constant. Teeth formed according to this principle have the following practical advantages over any other form of tooth:

1. All wheels with involute teeth will, if their pitch is the same, work correctly with one another.

2. They will also work correctly whether their centres are nearer or farther apart, this only influencing the duration of contact.

3. Involute teeth cause the least friction, and their original form is least changed by wear.

4. They are geometrically similar in form, and are therefore the most easily and most correctly to be cut by machinery. For these reasons involute teeth ought to be substituted for every other form of tooth.

In conical, or bevil wheels, the radii of the pitch circles must not be taken on the faces *a*, *b*, Fig. 7, but on the sides *c d* and *c e* of two supplementary cones, which sides are at right angles to the sides of the base cones *c f*. The form of the teeth is obtained for the largest diameter in the same manner as for spur wheels, and the bounding surfaces made to converge to the apex of the cones.

95. Having determined the correct geometrical form for the teeth of wheels, we have yet to consider the following points in their construction:

1. The necessary height of the teeth, for the duration of

contact to be as long as possible, and the proportion of the parts above and below pitch line.

2. The necessary thickness of the teeth and the proportion of the same to the space between them.

The height of a tooth is divided, as we have seen, into two parts, of different formation, the one termed the face being above the pitch line, the other termed the flank being the portion below the pitch line. These two heights are not equal, the portion below the pitch line being somewhat deeper, to allow a certain amount of play between the top of the tooth on the one wheel and the bottom of the tooth on the other wheel. The total height of the tooth is a certain function of the *pitch*, and is generally so arranged that at least three pairs of teeth are always in contact. The *pitch* is again dependant on the necessary strength or thickness of the tooth, which has to be determined according to the pressure which the driving wheel has to transmit. The *pitch* comprises the thickness of a tooth and the space for one of the other wheels to work into. Here also a certain amount of play is given to allow the teeth to clear well. In Fig. 8 are shown the generally adopted proportions of the heights and thickness of teeth and the spaces between them, where, the pitch being divided into 15 parts, the thickness is made equal to 7 of these, while the space takes 8; the total height of the tooth is made equal to twelve of these parts, of which $6\frac{1}{2}$ are allotted to the flank and $5\frac{1}{2}$ to the face.

96. Of late, millwrights have designated by *pitch* a different value to the one just contemplated, and have distinguished it from the latter by the term *pitch, new style,* while the other is called *pitch, old style.* The object of its introduction was the simplification of the relation between the diameter of the wheel, the number of teeth and their thickness, as it expresses by a fractional part of the diameter of the wheel the number of teeth for the same; or, in other words, suppose the diameter of a wheel to be divided into

as many equal parts as it has teeth, then one of these parts, expressed in inches or the fractional part of an inch, is termed *pitch, new style.* As the diameter multiplied with 3·1416 gives the circumference of the wheel, therefore the *pitch N. S.* being the portion of the diameter due to one tooth and a space, if multiplied with 3·1416, will give *pitch O. S.* For the sake of simplicity, $1\frac{1}{2}$ × pitch *N. S.* is taken for the thickness of the tooth, while the additional ·1416 besides the other $1\frac{1}{2}$ is allotted to the space for play. The total height of the tooth is made 2·1 times pitch *N. S.* and the height from bottom of tooth to pitch line is 1·2 × pitch *N. S.* By means of this modulus we know at once for instance that a wheel 2 feet diameter, $\frac{1}{2}$ inch pitch *N. S.* has 48 teeth, and that the teeth are $\frac{3}{4}$ inch thick. Fig. 9 shows the different values relatively to pitch *N. S.*, and Fig. 10 gives a full-sized diagram up to 4 inches pitch *O. S.* and $1\frac{1}{4}$ inch pitch *N. S.* of the thickness and height of teeth, P being = pitch *O. S.* and P′ pitch *N. S.*

97. As regards the strength of teeth, we can only take the thickness of the same into account, and not the width, as, though if accurately constructed, the contact between two teeth would take place along the whole width, yet we must suppose the worst case (as we cannot always insure perfect action), and that would be, when the corner of one tooth presses upon the corner of the other. In this case we have therefore to consider the tooth acting as a canter lever, whose length l (Fig. 11), is equal to half the breadth, b; as the proportion of l to b is evidently a maximum, when b is taken at an angle of 45°. The depth of the canter lever is the thickness t of the tooth, and we have the breaking strain for the same in lbs.:

$$\text{B. W.} = \frac{t^2 \times b \times 5600 \text{ lbs.}}{l}$$
$$= t^2 \times 11200.$$

98. In Plate VI. are shown the most usual constructions of spur- and bevil-wheels. In Figs. 1 and 2, *a* is what is

termed a *mortice-wheel*, that is, a wheel in which the teeth are of wood instead of iron, such an one being generally made to work with an iron wheel in mill gearing, as their action has been found to be much easier than that of two iron ones. The wooden cogs are of course made of greater thickness than the iron ones, for them to be of equal strength; they are fixed in their places either by a pin passing through them, or a dovetail-key driven between them, as shown. The iron wheel should always be the driver, and the mortice-wheel the follower. Fig. 3 represents the part plan, and Fig. 4 the section of a bevil-wheel; the lines of the teeth and rim are all made to converge to the apex *o* of the pitch cone, while the end surfaces are at right angles to the side of the same. These are also constructed as mortice-wheels.

Fig. 5 represents a pinion in gear with what is termed a *rack*. In this case the form of the teeth for the rack is that of the cycloid, the generating circle being half the diameter of the pitch circle, while the teeth of the pinion are formed according to the involute.

The rack changes the circular motion of the pinion into a rectilinear one, and may be considered as a spur-wheel of infinite diameter. It is, except when it forms part of a sliding framework or carriage, kept up to the pinion by means of a small guide roller at the back, furnished with a groove, into which fits a feather at the back of the rack.

99. The endless screw or worm, and worm-wheel, is often employed to transmit the circular motion of one shaft to another at right angles to it, but not intersecting the same. The nature of the motion is such, that for every revolution of the worm the worm-wheel performs a portion of a revolution corresponding to the pitch of the worm. Hence the speed of the screw shaft, when transmitted to the worm-wheel shaft, becomes very much reduced, and consequently the power very much increased. As regards the geometrical construction of the worm, it is formed in such a

manner, that if intersected by a plane passing through the axis, the section has the appearance of a double rack, the pitch of whose teeth corresponds with the pitch of the screw, supposing it has only one thread, or with the pitch of the threads, supposing it has more. The sectional form of the thread will also correspond with the form of the tooth of the rack, as shown in Fig. 6, Plate VI., and the teeth of the worm-wheel must be at an angle to the axis corresponding with the angle of the thread of the screw. As here represented, the contact between each tooth of the wheel and thread of the screw will, strictly speaking, only take place on a line formed by the intersection of a perpendicular plane passing through the axis of the screw. Where a more accurate contact is required, the form of the teeth for the pinion, at any vertical plane of intersection, must correspond with the section of the screw at the same point; and if it is desired to have a number of teeth in contact with the screw at the same time, the screw must be formed to fit the curve of the wheel, as in Fig. 12, Plate V.

The endless screw is sometimes adapted to work with the rack, in which case the teeth of the latter are at a corresponding angle to the pitch of the screw, as in Fig. 13, Plate V., or they may be formed in such a manner as to correspond with a portion of the thread of the female screw.

In all cases where the endless screw is employed, there is a great amount of power lost in the friction, resulting from the sliding action between the thread and the teeth. Where economy of power is an object, therefore, it should not be employed.

100. *Hyperbolical*—or *skew-bevil-wheels* are sometimes employed to transmit the motion from one shaft to another, where they are at an angle, but do not intersect. The most simple arrangement is shown in Figs. 7 and 8, Plate VI., where A is an ordinary bevil-wheel, and B the corresponding skew-bevil. The manner in which the teeth of the latter are formed is as follows:

With the smallest distance *c b* (Fig. 8) of the two axes, termed the excentricity, we describe the circle *b e d;* the line *a b*, being the direction of the axis of the bevil-wheel A, will be the line of contact for the teeth, and consequently also the direction of the face of the teeth for the skew-bevil. The other lines of the tooth are made to converge to this point, and the other teeth are formed by drawing the line of the working face as a tangent to the circle *b e d*. The line *a b*, from the contact of the pitch circles to the circumference of the circle *b e d*, if moved round the axis *c b*, Fig. 7, in such a manner as always to remain at the same angle and at the same distance, will describe a surface *f e* corresponding to the tangent cone of a hyperboloid; any portion of this tangent cone will work properly with a corresponding portion of the cone *g b h*.

101. *A variable velocity-ratio* may be obtained with wheel work, by a variety of contrivances, of which perhaps elliptical wheels (Fig. 14, Plate V.) are of most practical importance. Each axis *a* and *b* is fixed in one of the foci of the ellipse, and their distance *a b* is equal to the tranverse diameter of pitch line, the sum of the distances from centre of axis to point of contact being always a constant quantity. If the axis *a* revolves with constant angular velocity, then, as the smallest radius of the wheel A is in contact with the largest radius of the wheel B, and *vice versâ*, the velocity imparted to the axis *b* will be variable accordingly.

To obtain the generating circle for the form of the teeth, draw *a b* (Fig. 15), and *b c* perpendicular to *a b*, then *c d* will be the diameter for the generating circle *c d e* for all the teeth, which, though all of different form, will nevertheless act correctly. For a variety of other contrivances for varying the velocity-ratio, we must refer the reader to Professor Willis's "Principles of Mechanism," 2nd Edition, or to Baker's "Elements of Mechanism," numbers 98 and 98* in this series.

102. *A sudden change in velocity* may be obtained by an

arrangement as shown in Fig. 16, where the sectors A, B, C, D, on the shaft O revolving with constant velocity, come consecutively in contact with the sectors *a*, *b*, *c*, *d*, on the shaft *o*, and thus impart to the latter a velocity varying suddenly from that of O as the proportion of *a* : A; *b* : B, &c. These sectors are not in the same plane, but fixed one behind the other, and it is clear that the number of teeth in each sector on O must be equal to the number in each *corresponding* sector on *o*. It would carry us beyond the limits of this little work were we further to enumerate separately the inexhaustible stock of mechanical contrivances for modifying motion, and we must content ourselves therefore with a description of such as may occur in the machines treated upon in the next chapter, referring the reader to the above-named works for further information on the subject.

CHAPTER VII.

OF VARIOUS DESCRIPTIONS OF MACHINES CONSTRUCTED FOR DIFFERENT PROCESSES OF MANUFACTURE.

HAVING investigated thus far the leading features common to all machinery, we have now to consider their combination in certain machines, carrying out different processes of manufacture.

SECTION I.

103. *Machinery for the Manufacture of Iron.*—The various machines employed in the process of converting cast into malleable iron, as also those for shaping and moulding the latter, have all two features in common, namely:—1. Great strength in the different parts; and 2. Simplicity in their action. Those machines, to the consideration of which we here confine ourselves, simply carry out a process of squeezing, forging, rolling, shearing, or punching, the object of which operations we will now shortly explain.

104. Cast iron, as produced from the ores in the blast furnace, contains beside other foreign elements, large quantities of carbon and silicon. It is the removal of these from the ore which constitutes the conversion of cast into wrought iron. This is carried out, first, by a process termed "refining," which consists in bringing the cast iron in a molten state in contact with a blast of air, the oxygen of which combines with the carbon and silicon of the iron,

causing the former to pass off in the form of gas, and the latter in the form of a siliceous slag. Secondly, the further decarbonisation is carried on in a reverberatory furnace by a process termed puddling, by which the iron loses a still greater portion of the carbon, and is consequently no longer in a fusible state, but assumes a plastic form. This *puddling* consists in continually stirring the iron about with a paddle, and eventually, when it has become pasty, forming it into *balls* of from 60 to 70lbs. weight each. These balls are then removed from the furnace, and placed either under the hammer, the squeezer, or, as in Wales, directly under the roughing rollers, by which various operations the slag is expressed, and the spongy *ball* converted into a solid mass of wrought iron.

105. The ordinary tilt hammer, as shown in Fig. 17, Plate V., consists of a cast iron helve A, weighing about four tons, and about 10 feet long, which is supported at the end B by massive plummer-blocks, fixed upon wooden beams to take up the powerful vibrations. The head C is of wrought iron faced with steel, and is fixed to the helve by passing through a hole in the same, and being firmly keyed in. The part D of the anvil is also of cast iron, while the *pane* E is of wrought iron, and faced with steel, like the *head.* The meeting surfaces are grooved for better expressing the slags. The head is raised by means of a succession of cams F F, fixed in a massive cast iron collar G upon the cam-shaft H. These raise the hammer from 16 to 24 inches high, causing the same to make from 75 to 100 blows a minute. The collar G termed the *cam-ring-bag* is about 3 feet diameter, and 18 or 20 inches thick, weighing from 4 to 4½ tons. When the hammer is not in use, it is propped up by the support I. The cam-shaft is provided with a heavy fly-wheel.

106. Of late years, this hammer has in a great measure been superseded by Nasmyth's *Steam Hammer*, for a full description of which, by Mr. Nasmyth himself, together

with excellent drawings of the same, we must refer the reader to "The Engineer and Machinist's Assistant." We give in Plate VII., Figs. 7—10, a modification of the same, as constructed by Cavé, of Paris. The hammer-block A is of cast-iron, of which the cross section is shown in Fig. 10; the hammer itself, B, is fixed to this by means of the dovetailed portion, which is keyed in. This hammer-block is guided accurately perpendicular between the guides D, fixed to the uprights C, of the frame-work, by means of the bolts and keys, *b*. The uprights are connected by means of the casting E, to which they are fixed by means of the wedges *c*. On the top of this casting is fixed the steam cylinder F, of a length equal to the greatest lift required to give the hammer. The top of this cylinder is closed by the cap G, which is removed when the packing of the piston H requires renewing. This piston consists of two moulded wrought-iron discs, fixed together, and also to the piston-rod *d*, by the end of the latter passing through them, with a shoulder on one side, and a nut on the other. M. Cavé prefers hemp to metallic packings for the cylinders, as it does not cause so much wear and tear. The piston-rod *d* passes through the gland of the cylinder, and is connected directly with the hammer-block A, in such a manner that there is a packing of hemp or wood between the two to allow a certain amount of elasticity. The steam is admitted through the port *i*, and passes out through the exhaust port, *i'*. The slide-valve box I is provided with a small cylinder J, in which works the piston K, connected by means of the link *j* to the slide valve *j'*. The object of this arrangement is self-evident: the steam pressing both upon the slide-valve and upon the small piston, the area of which is rather less than that of the former; the actual force pressing the slide-valve against the face of the ports is thus greatly reduced, and the slide valve is thereby rendered easily manageable by hand. The slide-valve rod K is connected with the double lever L, and this receives its motion through the rod M from the lever N, which is worked by hand. The operator who works the valve stands upon the small raised platform P. Another lever Q serves to check the downward course of the hammer by means of a small nosing, fitting into a corresponding notch in the hammer-block. The anvil S is fixed into the anvil-block T, in such a manner that it can be exchanged at pleasure; this latter is a heavy mass of cast-iron, so as to withstand, by means of its inertia, the shock of the blows.

The upper portion of the cylinder is in communication with the atmosphere, by means of the two small valves *r* and *s*, the former allowing the air to pass out of the cylinder as the piston moves up, while the latter allows the air to re-enter the cylinder when the piston is falling. As soon as the piston passes the valve *r*, it has to compress the air still contained in the upper part, which thus forms a cushion, preventing the piston from flying up against the top of the cylinder.

107. *Friction hammers* are also applied to similar purposes; they derive their name from the hammer being lifted up by means of the friction of revolving rollers, nipping a rod in connection with the hammer. Figs. 8—11, Plate VIII., represent a hammer of this description, as constructed by Mr. Eassie, of Gloucester. The frame-work consists of a pillar A, which is cast in one piece with the foundation plate B, and to which are bolted the two cheeks C C, carrying the working apparatus. The centre part D is provided with bearings, in which the strong wrought-iron axle, carrying the oscillating frame E E, works. On this frame are fixed the plummer-blocks for the spindles, carrying the driving gear and friction rollers. The motion is, in the first instance, communicated to the drum G, and, consequently to G′, fixed upon the same shaft; this communicates the motion by means of a crossed strap to the drum H, on the shaft H′, and hence also to the pinion K, which

works into the spur-wheel L, fixed upon the same shaft with the friction roller N. In a similar manner, the friction roller N′, is set in motion by means of the spur-wheel L′, the pinion K′, and the drum G′.

The friction rollers are provided on their circumference with a wooden rim, and, in the event of their wearing, may be adjusted by means of the set screws X X. The principal feature in this contrivance consists of the oscillating frame E, which, when tilted upon its axis, causes the friction rollers N N′, to nip the hammer-rod, the one a little above, the other a little below, the axis. When the frame E resumes its horizontal position, the hammer is allowed to fall, being guided by the rollers P P and Q Q. The tilting of the frame is done by means of the lever M, connected with the same by means of the rod and chain passing over the fixed pulley J.

The hammer and rod of this apparatus weighs about 20 cwt., and gives from 100 to 150 blows a minute, with a lift of about one foot.

108. The *squeezer*, or hinged press, is often preferred to the hammer for working the balls of puddled iron, as they do not occasion so much waste as the latter. Figs. 1—6 Plate VII., represent a description of this apparatus as constructed by M. Cavé, in which, as in the steam-hammer, the direct action of the motive power is applied. This is a most judicious arrangement in an economical point of view, and should be, where possible, adapted to all machines requiring considerable power, which do not work constantly. A A′ is a massive lever of cast-iron, provided with a boss in the centre to receive the axle B, upon which it swings. At the end A it is forked, and provided with two smaller bosses, one for receiving the cross-head of the piston rod of the steam cylinder which actuates the machine, the other for the spindle of the link C, turning the crank shaft H, upon which two fly-wheels K K are fixed; from this link C the slide-valve is also worked by means of the levers

m, *l*, and *l'*. The working end of the lever A' is provided with thick cast-iron plates *a a*, for protection, which are fixed to the same by means of the bolts with countersunk heads, as shown in the section at *b b*, Fig. 5.

The steam-cylinder Q is open to the atmosphere at the top, being only single acting. The piston P is of considerable depth, having both a metallic and a hemp-packing, and is connected with the piston-rod in such a manner that the latter can vibrate in a corresponding manner with the end of the lever A, which naturally describes an arc in its upward and downward motion. The framework carrying both the lever and the steam-cylinder is likewise of cast-iron and of two pieces, being divided longitudinally and bolted together at the bottom. It is provided with gun metal bushes at F, which receive the axle B, and at E forms the anvil upon which the iron is worked. The plates D are separate castings and of considerable thickness to withstand the pressure, while the part E is full of water to prevent their becoming overheated; they are so arranged that they can be replaced when worn out. This machine requires no particular foundations, as the whole is carried by the framework E, and can therefore be placed close to the puddling furnace, which is a great convenience.

The total weight of this squeezer is about 250 cwt., and it makes from 40 to 60 strokes a minute. Its cost is much less than a steam hammer of equal efficacy.

109. After the puddled ball has been under the operation of either the hammer or the squeezer, it is further subjected to the process of *rolling* in the *rolling mill*. These are of two kinds: the *roughing*, or *pudding roller*, serving much the same purpose as the hammer and squeezer, namely, by compressing and elongating the puddled iron, giving it greater tenacity. After once being passed through these rollers the bar is generally cut up in pieces, which are placed side by side, welded together and again passed through the finishing *rolling mill*, which consists of a pair of massive

rollers furnished with grooves, corresponding with the form of the section which the bar iron has to receive. The grooves for the *roughing* and *finishing* are frequently on the same rollers, but generally they are on distinct rollers. Figs. 1, 2, Plate VIII. represent a rolling mill, where *a a a* are roughing grooves, while *b b b* are finishing grooves for square iron; *c c c* are separate rollers for flat iron.

The roughing grooves *a a a* are of an elliptical form and of a gradual decreasing size, being so arranged that the shorter axis of the ellipse of the first pair of grooves is equal to the long axis of the second, and so on. The small axis being in each case perpendicular to the axis of the rollers, it follows that the bar of iron is turned 90° on its axis when passing from one groove to another, so that the same is equally compressed and drawn out on all sides.

The rolling mill here shown has three rollers, by which the workman is enabled, when the bar has passed through one groove at the bottom, to pass it back through the next sized groove at the top, thus saving the time necessary to pass the iron back to effect this, which has to be done where there is only a pair of rollers.

In the finishing grooves for square iron *b b*, the diagonal of the square is parallel with the axis of the rollers, which enables the bar to clear the grooves more readily than if the side of the square were parallel with the same.

The baseplate A A is fixed upon a very solid foundation, and generally has, immediately beneath it, a strong framework of timber which serves to take up the violent vibrations resulting from the shock of the metal entering the rollers. In this baseplate are fixed, by bolts and folding wedges for adjustment, the uprights or *housing frames* B, which form the bearings for the roller gudgeons to run in. These gudgeons are of one casting with the rollers, and are adjusted by means of the screws C C, the blocks D D, and the wedges E.

The motion is communicated to the rollers from the shaft

G, which is connected with the steam-engine or water-wheel, by means of the strong spur wheels F F. The couplings H H H are formed in such a manner that, should by mistake the *mill bar* be forced into the wrong groove, these would give way, and thus prevent the rollers from breaking.

The *mill bars* are supported in front of the rollers by means of a strong iron plate termed the *apron*, which rests upon iron bars running from frame to frame and fixed in the grooves I I.

110. A similar description of rollers are employed for making sheet iron, having smoothed and polished surfaces. Square bar iron of very small scantlings is formed out of sheet iron by means of a description of rollers termed *slitters*, as shown in Fig. 18, Plate V. They are a series of steel discs with very sharp edges, with a groove between them of about 2 inches in depth, into which the discs of the other roller fit. A piece of sheet iron being passed through the guides *a* is divided into a number of strips corresponding to the number of grooves.

111. Fig. 19, Plate V., represents the *shears* ordinarily employed for cutting up the *bloom*, or mill bar, as it comes from the roughing-rollers. The cutting edges *a b* are of hard steel, and are firmly bolted, one to the fixed frame *d*, the other to the massive lever *c*, which has a boss *e* to receive the axle *f*, upon which it hinges, and at the lever end *g* receives the end of the connecting rod *h*, moved by a crank from the fly-wheel shaft.

112. Another description of shearing-machine, which is also provided with a punching apparatus, is shown at Figs. 11, 12, Plate VII. This apparatus is, however, not used for this kind of work, but rather in the workshop of the engineer. The cast-iron frame A has bolted to it at C one of the steel-cutting edges, which here obtain a slanting form, so that the iron is not cut through the whole section at once, which would require too great a sudden

force, and would not make so clean a cut. Below, this frame carries the punching die B, adjusted by means of the set screws. It also carries the spindle D, upon which are fixed the fast and loose drum Q Q, the fly-wheel F, and the pinion G. This pinion works into the spur-wheel H, and thus turns the spindle E, which at the front end I is formed excentrically to the amount of half the stroke of the slide L, carrying the other portion of the shears N, and the punch O. It imparts its motion to this through a species of toggle joint K K. The slide L works between the cheeks M M, adjustable by means of set screws. P is a guard for removing the plate, which may adhere to the punch after the hole is punched.

113. Figs. 3—7 represent a machine for forging small articles invented by Mr. Ryder, which has been very generally introduced in all larger workshops on account of its great efficacy.

The cast-iron frame A carries at the top the shaft B, upon which are formed several excentrics *c c c*, acting upon the top of the die spindles D, furnished on this account with a brass saddle. These spindles are guided very accurately in gun-metal bushes E E, fixed in the cast-iron frame F. They are hollow, and contain a spiral spring, pressing at the top against the spindle itself, and at the bottom against the key *b*, passing through a slot in the spindle, and fixed in the frame F. By this arrangement, as soon as the excentric begins to recede after having depressed the die spindle, this latter is pressed up against it by means of the spring which has been compressed. The dies G, I, and chisel H, can be replaced by whatever form may be required. The motion imparted to these dies is exceedingly rapid, the excentric shaft C, upon which is fixed the fast and loose drum P P′, and the two small fly-wheels O O, making from 600 to 700 revolutions per minute.

The bottom dies I I, J J, are not perfectly rigidly fixed,

but have inside either a steel or india-rubber spring, against which screw K presses, and by turning this screw, the whole of the bottom die, which also slides in a gun-metal bush *h h*, fixed in the frame L, may be raised or lowered, and thus the intensity of the pressure, exerted by the top die, modified. With the chisel this is different; here, instead of a screw, a small excentric is fixed upon the spindle M, which is turned by means of the lever N. The excentric shaft B is carried at each end by the bearings Q Q, and also between each excentric by the bearings R R. The bracket S carries a moveable rest O, which serves as a gauge for the portion to be forged or cut off.

114. In conclusion, the following may give an idea of the motive power required for the different machines for manufacturing wrought-iron.

At the *Dowlais Iron Works, in South Wales*, the following horse-power are required for working the different machinery:—

	HP.
4 Pumping Engines	496
16 Lifts, &c.	1136
14 Feeding Machines to Blast Furnaces	668
11 Locomotives	580
2 Machines for Pottery, &c.	76
5 Blast Engines	2154
10 Rolling and Forging Mills	2165
2 Turning and Boring Engines	35
Total	7310

Yearly production of iron . . . . . 108,576 tons.

2 rolling-mills, 1 double-squeezer, and 2 shears, when producing 300 tons of bar iron, 3 in. × $\frac{3}{4}$ in. scantling weekly, consume 74 HP, viz.:—

	HP.
Friction of steam engine	7
Ditto of heavy shafting, &c.	15
Shears	4
Rolling mills	37
Squeezers	14

Total H.P. required for making 600 tons of flange rails weekly, 239 H.P., viz.:—

	H.P.
Friction of Engine	12
Ditto heavy shafting, &c.	22
Rolling-mill	179
Shears	6
Eight straightening presses	12
Circular saws	8

The 179 H.P. for rolling-mill is the mean power which is accumulated in the fly-wheel, the actual power required while rolling being from 332 to 485 H.P.

A 12-inch rolling-mill, with horizontal high-pressure engine, at 140 revolutions, when producing weekly 100 tons of bar-iron, 1 in. × $\frac{5}{8}$ in. scantling, 49 H.P.

An 8-inch rolling-mill for $\frac{3}{4}$-inch flat iron, at 280 revolutions, 62 H.P.

Shears for bar-iron, 6 ins. × 1 in., 65 cuts per minute, 9 H.P.

SECTION II.

115. *Machinery for the Manufacture of Cotton.*—The peculiar features of this class of machinery are directly the reverse of those appertaining to the class we have just considered; for, whereas in the latter we had the elements of strength and simplicity of construction embodied, we shall find in the machinery now under consideration, delicacy of the working parts, together with the most refined, and in some cases intricate, mechanism, to be the prevailing features.

116. It need hardly be stated that cotton, in its raw state, consists of an entangled mass of vegetable fibre, contained in the seed-pod of the cotton plant. The process of manufacture may be considered to commence with the separation of the cotton-wool from the pod and seeds. This is effected by means of an apparatus as represented in Fig. 1, Plate IX., consisting of from 50 to 60 small sharp circular saws *m*, fixed

on the spindle *n*, of about 12 inches diameter, which are kept a small distance apart by thin wooden washers. These circular saws are covered in such a manner by the curved iron grating *o*, that the former just project through, one between each two bars of the grating. This grating forms part of the hopper *q*, into which the cotton to be cleared of seed is thrown. The portion *r* of this hopper is adjustable by means of the screw *s*. At the back of the circular saws is placed a drum *t*, upon which a series of brushes are fixed; these revolving in the same direction as, but with a greater velocity than, the saws, serve to clear the very acutely formed teeth of the latter, of the cotton hanging to them. The saws revolve 100 times, while the brushes revolve 150 times, a minute. The former, projecting into the hopper, catch hold of the cotton fibre, dragging the same through the grating, whilst the seeds, too large to pass through, remain in the hopper, and fall out of the same at *r*. This separating process is generally carried out at the plantation, the cotton is then closely packed—now generally by hydraulic pressure —for exportation.

117. The actual manufacturing process by which a continuous, fine, twisted filament is formed out of this compressed and entangled mass of fibres, may be divided into the six following partial processes:—

1. The *loosening* and *cleaning* process, by which the compressed cotton is imported, as opened up and separated from impurities.

2. The *carding* process, by which the cotton fibres are disentangled and placed in a parallel position to one another.

3. The *drawing* and *doubling* process, actuating a more complete parallelism of the fibres, and equalising the texture of the *sliver* or *band*.

4. The *roving* process, by which the *sliver* is very much *drawn*, and receives a slight twist to give it sufficient strength.

5. The *fine roving*, or preliminary spinning process, giving a considerable degree of fineness and twist to the thread, and preparing it for

6. The *spinning* process, completing the thread for weaving purposes.

We have now to consider, separately, the machinery necessary for carrying out each of these processes.

118. The *loosening* and *cleaning* operation is carried out—

(1.) By the *opening machine*, as shown in Fig. 2, Plate IX., which is principally employed where the *water-twist frame* is used.

It consists of a wooden drum *a*, on the circumference of which are fixed rows of iron spikes alternately with the upright iron ridges *c c c*, which ridges serve to prevent the cotton from passing too rapidly through the machine. This drum is covered at the top with the wooden lid *d d*, consisting of strips of wood curved to the shape of the drum, and furnished with spikes and ridges similar to the same. These strips are so fixed to the frame-work, that they are singly removable for cleansing or repairing purposes. The lower part of the drum is covered in with the brass wire gauze *r*, furnished with an opening at *r'* through which the cleaned cotton is removed, while the dust and other impurities fall through the gauze. The machine is fed from the *feed cloth* *p*, upon which the cotton is spread out by hand, and from which it is taken by the grooved nipping rollers *k l*. The drum is driven by means of the pulley *b*, at the rate of from 400—450 revolutions per minute. This machine can prepare from 1000 to 1200lbs. of cotton a day.

(2.) The *willow* is a similar machine to the one just described, being of a coarse nature, as it is employed for cotton of a long fibre, and containing a great deal of impurity. Figs. 3 and 4, Plate IX., represent a sectional elevation and plan of the same. It will be seen that the spikes *c f* are less numerous, but much longer, those on the drum passing between those on the cover. The machine is fed through the hopper *p*, and the cotton is taken out through the door *r*, while the impurities pass through the grid *h*. The drum, which is driven by means of the fast and loose pulleys D at the rate of 300 revolutions per minute, can prepare 150—200lbs. of cotton per hour. Each charge consists of about 1¼ pound, and the duration of each operation about 25 seconds.

(3.) The *batting machine, blower* or *scutcher*, as shown in Fig. 1, Plate X., serves to separate each individual fibre, after the cotton has been cleansed in the machines just described. This machine is fed by means of the cotton being spread by hand on the feed-cloth *a a*, passing over the rollers *b c*, and moving slowly in the direction of the arrows. The roller *c*, which is 3 inches in diameter, is driven at the rate of four revolutions per minute. The cloth is supported by the table *d*, and consists of a series of bands, which can be strained more uniformly than would be the case if the whole were one cloth. At *c* the cotton is caught between the nipping rollers *e*, which present it to the batting arm *f*, revolving 640 times a minute, and consisting of the arms *20*, at the end of which are fixed, longitudinally, two narrow thin strips of sheet iron. These, catching the cotton presented to them by the nipping rollers, tear it off, separating the fibres from each other, and letting the dust, &c., fall through the grid *n*. The fibres thus torn asunder are again collected upon a second feed-cloth *a'*, where they are pressed together by the drum *h*, and then again caught by the nipping rollers *e' r'*, to subject the cotton to a second batting operation at *f'*. As only the heavy impurities pass through the grid *n*, a sucking action is provided by means of the fan *p'* over the perforated compressing drum, revolving 150 times per minute, by which the light dust is sucked up and conveyed through the channel *12* into the box *13*, where it is deposited. From the second batting apparatus the cotton fibres are thrown into the closed channel *x*, and are driven through the same by a second fan *m q*, revolving 700 times per minute. At *w* is another grating, consisting of bars slantingly placed, and at *n v* are two doors through which the cotton is taken out. The method of driving the different apparatus is shown in Fig. 1 for one side, and in Fig. 2 for the other. In Fig. 2 are shown the lever *14* and weight *15*, by which the nipping rollers are pressed together.

This batting machine can produce 500 to 600 lbs. of scutched cotton in 12 hours.

(4.) The *spreading, or lap machine*, is in its first operation similar to the scutcher, but derives its different name from an apparatus attached to it, by means of which the cotton, after having been subjected to the batting process, is formed into a continuous band, preparing it for the carding process.

Fig. 5, Plate IX., represents the same, where *a* is the feed cloth as before, *e e* are the nipping rollers, *f* is the batting arm, *n* the grid, *h* the perforated pressing drum, and *p p* the suction fan. After having been compressed on the second feed cloth *q*, the cotton passes between the smooth iron rollers *r s*, which are pressed together by means of heavy weights, and is then wound up in a continuous band upon the roller *v*, which is again kept firmly pressed against the two wooden rollers *t u*, by the weight *x*; these two rollers are made to revolve with a speed corresponding to the velocity with which the band is delivered; these turn the roller *v* by friction, the axis of which, as it increases in size as the lap is wound round it, move up the slits at each side of the machine.

When a sufficient quantity is wound upon the roller, the weights *x* are lifted off the same by the lever *z*, and are suspended to the hook *y*.

As the character of the cotton thread already depends to a certain extent upon the uniformity of the lap produced by this machine, great care has to be taken with regard to the feeding of the same. For this reason the feed cloth is divided into equal parts by red stripes, and upon each division a certain weighed quantity is spread very carefully. When distinct lengths of fleece are to be produced, one of these divisions are left uncovered. The spreading machine is of the same width as the carding machine, namely, 3 feet, and can produce in 4½ minutes a fleece of 30 feet in length, and 5 pounds weight; or, in 12 hours, 650 pounds of cotton fleece.

119. In the *carding process*, the cotton fleece, as it comes from the spreading machine, undergoes a combing action, by means of what are termed cards. These are formed in the first instance of long narrow strips of leather *a b*, Figs. 6 and 8, Plate IX., which are pierced with holes for the reception of bent pieces of steel wire, of the form as shown in Fig. 7.

If two such card surfaces be brought in contact in the position as at Fig. 6, with some cotton-wool between them, and be moved in the direction as shown by the arrows, then the fibres will be drawn out of their entangled and bent position, and by the continued action of the surfaces, will eventually be laid perfectly parallel, partly on one surface, and partly upon the other. If, after this is effected, it is required to clear one card of the fibre, it is necessary to reverse the position of the two surfaces, as in Fig. 8, when the surface *a* moving over *b* will strip the latter clear of fibres.

The operation of carding is carried out by two machines, first by the *breaker*, and then by the *finisher*. The first named is similar to the *opening machine* before described, consisting also mainly of a feed cloth, a drum, on the circumference of which the just mentioned card band is wound, the card being of a coarse description. This drum is covered in at the top with a card surface, placed as at Fig. 6; and at the side opposite the feed cloth, is in contact with another smaller card drum, revolving in the reverse direction to the large drum, so that the surfaces in contact are moving in the same direction, though that of the smaller drum moves with a less linear velocity than the large one, and thus clears the latter of the carded cotton. From this small drum a thin fleece is taken off, by means of an apparatus which we shall describe in the *finishing card*. Several of these fleeces are joined to form one thickness by the *lapping machine*, as shown in Fig. 9, Plate IX., where several cans *o o*, filled with a narrow fleece from the breaker card, are placed in two rows, and the ends of the fleece brought together between two rollers, the lower one of which is driven by the wheel and pinion *l* and *b*, and the upper one *a* pressed down upon the former by weights.

After being joined together by this pressure, the fleece is conducted over a hinged table *e*, which, by means of the wheel *k* and the levers *j* and *g*, is moved slowly up and down, as shown by the dotted lines, while the fleece, passing slowly over the same, is folded up and down in the box *p*. Several of these boxes are again brought before a second lapping machine, which is precisely the same in its construction as the spreading apparatus connected with the second batting machine, the fleece being rolled upon rollers ready for the *finishing card*.

The *finishing card*, Fig. 10, Plate IX., receives the roller from the lapping machine in the grooves *a*, resting at the same time upon the roller *b*, which turns it by friction, so as to unwind the fleece or *lap*. This is conducted over the table *c*, between the two nipping or feed rollers *e* (1·16 inch diameter, 0·69 revs. per minute), where it is caught by the card drum *f* (35 inches diameter, 130 revs. per minute), covered with the card cloth in strips, parallel with its axis; *h'* is the small runner (3·5 inches diameter, 470 revs.); *i'* the large runner (6·25 inches diameter, 5 revs.); *g g g* the flat top cards; *h* the *doffer card* (14 inches diameter, 4·38 revs.), which takes the fleece off the main drum, and which in its turn is cleared by the *doffer knife*. This knife receives a very fast up-and-down motion from the crank *m*, by which it peels, as it were, a gossamer-like fleece from the doffer card. It is actually a comb, with very fine teeth, which penetrate slightly between the wire of the card as it moves downwards. The fleece thus separated from the card is then drawn together through the tin funnel *n*, so as to form a narrow band or *sliver*, and is thus presented to the *drawing rollers* at *o*. The front pair of these (1·16 inch diameter, 114·5 revs.), being driven with a slightly greater velocity than those behind (1 inch diameter, 68·71 revs.), draw the sliver through faster than it is delivered by the latter, and by this means extenuate the same. This action brings about a still greater parallelism of the fibres, and we shall find it repeated in every subsequent machine. The bottom rollers are of iron and fluted, and are driven by gearing; the top ones are covered with leather, and are pressed down upon the lower ones by the weight *w*. The rollers *u v* deliver the drawn sliver into cans, ready for the drawing and doubling process.

It will be seen from the relative speed of the main card drum *f*, the large runner *i'*, and the small runner *h'*, which is respectively 14,287 inches per minute, 98·1 inches, and 5170 inches, that the large runner will take the cotton from the main drum *f*, the small runner will then take it from the large runner, and this will deliver it on to the main drum again.

120. The *drawing* and *doubling process* first draws out the slivers as produced by the finishing card by means of drawing rollers, and then unites several of these into one. The object of the first operation is, as already stated, to draw each fibre past the next one, thus placing them still more completely parallel to each other, while that of the second is to neutralise the inequalities in each separate sliver, and to strengthen them after having been extenuated.

Figs. 11, 12, Plate IX., represent the *drawing frame*, while Fig. 5, Plate X., shows an enlarged section of the drawing rollers, which are the same for all the subsequent machines. The cans from the finishing card are placed at A and the slivers led over the table *s* through the fork *t* to the drawing rollers. Of these, as already stated, the second pair revolve rather faster than the first, and the third pair rather faster than the second. The increase of the velocity is obtained both by the increase of the diameter and an increase of the number of revolutions. After having been drawn out, the ribband passes through the tin funnel *m*, and through the compressing rollers *p*, and is again delivered into a can. These cans are again put before a second similar machine, where several slivers are united and drawn out again, which operation is repeated four times; after this the cans are brought to

121. The *can roving machine.* This machine, Fig. 13, Plate IX., is provided with a drawing apparatus *a b*, like the drawing frame, but the cans into which the sliver is delivered, instead of being stationary, are fixed to a pivot at the bottom and supported at the neck *f*, while they are made to revolve by a strap passing round the pulleys *n* and *k*. This rotatory motion gives the sliver a slight twist as it passes in, and coils it up in the can in a regular manner, the twist being necessary to prevent the sliver tearing off during the subsequent operation, it being in a very attenuated state.

When the can is full it is opened, the *roving* taken out and transported to a roving machine, where it is wound upon reels ready for spinning.

The fact that this last operation almost always spoils the delicate sliver to a certain extent, has been the cause of this otherwise perfect machine having been generally abandoned.

122. The *bobbin and fly frame* has been generally substituted for the can roving machine. This machine is, as regards both the ingeniousness of its conception and the precision with which the different parts work, one of the most perfect specimens of automaton machinery existing.

Fig. 3, Plate X., represents the back elevation of this

machine, that is, the side where the cans from the drawing frame are placed. Fig. 4 is a tranverse section, showing such parts as are not quite distinctly visible in the elevation.

The slivers are led from the cans up the table *f* through the fork *e* to the drawing rollers *a b*, of which there are three pairs, as in the drawing frame; *g g* are the weights for pressing the top rollers on to the lower ones. In the elevation the greater part of both the table and the drawing rollers have been omitted for the sake of clearness. After having been sufficiently drawn out, the sliver, now of a very delicate texture, is led down to the spindles *i i*, where it receives a certain amount of twist and is wound upon the bobbin *h*.

The drawing apparatus is supported from the framework A by a number of bearers *c*. The grooved rollers are formed four in one piece, with a bearing between each, and the different sets of four are coupled together by the last roller, having a square hole in it, into which the square end of the next set fits. The top rollers are formed two and two together.

The construction of the spindles may be seen more clearly in the enlarged view of the same, Fig. 7, Plate X. They are of iron, and made cylindrical from the top to a^2, from where down to the bottom they are conical, having at the end a steel pivot. Upon the conical part is fixed a pulley *k*, with two grooves, by which the spindle receives its rotatory motion. The bobbin *h* fits upon the cylindrical portion in such a manner that it can slide up and down it easily. To the bobbin is fixed another pulley with two grooves *q*, by means of the pin *r*, which is removed when the bobbin has to be taken out. Upon the top of the spindle is fixed the fork *s t*, which is also easily removable. It has a small funnel at *v*, with a hole in the side of it. The leg *s* is tubular, and the other, *t*, is only to keep the equilibrium. The spindles are ranged in two rows in a zigzag upon the framework *m*.

The sliver from the stretching rollers passes through

the hole at *v*, and down the tube *s*. If the spindle and the bobbin, which revolve in the same direction, went at the same speed, the only effect would be to twist the sliver, but as the spindle moves rather faster than the bobbin, the twisted sliver is at the same time wound round the bobbin. The manner in which the spindles, with the fly and the bobbins, receive their motion is as follows:—

A spindle c^1 provided at the one end with a fast and loose pulley a^1 b^1, and at the other with a fly-wheel d^1, runs through the whole length of the machine, and is driven from the main shaft of the mill. It has further fixed upon it the spur-wheel b^2, which gives motion by means of the spur-wheels c^2 d^2 e^2 f^2 to the drawing rollers. For imparting motion to the spindles, two pulleys e^1 f^1, are fixed upon c^1, each having four grooves, and each one having to move half the number of spindles (of which the machine here represented contains 30), by means of an endless gut, which passes four times round the same and the guide pulleys h^1 i^1, and then round the pulleys of the spindles, in the manner shown at Fig. 8, Plate X., where the numbers denote the order in which the gut passes round them. These spindles revolve at the rate of about 200 revolutions per minute.

Before describing the manner in which the bobbins receive their motion, it is necessary that we should explain the peculiar manner in which that motion must take place, in order that the cotton thread can be wound upon them in a proper manner.

Firstly. The cotton has to be wound up and down upon the bobbins, which, as the spindles and flies are fixed, as we have seen, can only be effected by the bobbins having alternately an upward and downward motion.

Secondly. As the cotton is wound upon the bobbin, the diameter of the latter naturally becomes gradually enlarged; and, if the speed of the bobbins were to remain constant, a great deal more cotton would be wound upon the same during an equal space of time, when the bobbin is getting full, than there would be when it was empty. But as the quantity delivered from the drawing rollers is always constant, this obviously can not be. Some contrivance, therefore, has to be adopted to decrease the velocity of the bobbins as they fill. For the same reason, the up-and-down motion of the bobbins has to be retarded as they fill. This up-and-down motion is effected in the following manner:—

The spindle c^3 receives its motion, in a manner to be hereafter described, from the spindle c^1, and has at one end a small pinion d^3, which works into the mangle wheel f^3, fixed upon the spindle e^3. The mangle wheel and pinion are shown more clearly in Fig. 6, Plate X., whence it will be perceived that when the pinion d^3 arrives at the opening g^3, it will turn over to the inside, as shown by the dotted position; and thus, if it continually move round in the same direction, it will alternately turn the mangle wheel first in one direction and then in the other. To the spindle e^3 are attached two pinions n^2, which work into and move alternately up and down the two racks m^2. These racks are fixed to the iron frame *p*, which consequently participates in their motion, being guided by the fixed framework, and balanced by the balance-weight m^4. Upon this movable frame *p* are fixed brass rings *o o*, serving the double purpose of guide to the spindles *i i*, and rest to the bobbins *h h*, which they cause to slide up and down upon the spindles.

Thus far we have shown how a constant up-and-down motion is produced, and it remains to explain the contrivance by which the same is rendered variable.

The long end of the bell crank t^1 v^1 is fixed in the slot *u u*, forming a portion of an apparatus to be hereafter described, which, as the bobbins fill, is caused to move gradually in the direction of the arrow. This has the effect of lifting the short lever v^1 of the bell crank, and with it the upright spindle y^2, having at the bottom the wide pinion w^1, and at the top the roller x^1, being guided at x^2. This spindle receives motion from the main shaft c^1, by the roller x^1 being kept firmly pressed against the revolving disc g^1, fixed on c^1, by means of the weight z^1. Now, as the spindle y^2 moves upwards, the roller x^1 is pushed radially upon the surface of the disc, which, therefore, gradually turns the same, and with it the spindle y^2 and pinion w^1, slower and slower. The pinion w^1 is in contact with the spur-wheel z^2, which turns the bevil wheel b^3, fixed upon the spindle c^3, on which is also the pinion in contact with the mangle wheel, which thus obtains the desired retarded motion. The spindle c^3 has a bearing at i^3, in the manner shown more clearly at Fig. 11, while at the other end it has a slight backward and forward motion, due to its change of position when inside and outside of the mangle wheel.

We have now to describe how the bobbins receive their gradually retarded motion.

The conical drum k^1 is connected, in a manner shown more clearly in Fig. 10, with the main shaft c^1, so that it receives its rotatory motion from, but is free to slide backwards and forwards upon, the same. The gut n^1 passes over this drum, and by means of the guide pulleys l^1, round the pulley m^1, fixed upon the upright spindle s^2. This spindle has another pulley, t^2, fixed upon it in such a manner that it turns with the same, but is free to move up and down it. From this pulley, which has four grooves, are driven the bobbins, in the manner as shown by the plan Fig. 9, where t^2 is the pulley, the figures denoting the order in which the gut passes round the same and the pulleys of the bobbins. The pulley t^2 is furnished with a neck, into which fits a fork u^2, fixed upon the movable frame p, and thus it is carried up and down in the same manner as the bobbins.

As the gut passes from the largest to the smallest diameter of the conical drum k^1, the shaft c^1 always revolving with a constant velocity, the pulley m^1, and consequently also t^2 and the bobbins, must obviously gradually decrease in speed. The gut n^1 is kept taut by means of a weight o^1 attached to the guide pulleys l^1, which slides up and down the rod p^1.

There now only remains to be shown how the conical drum receives its motion along the shaft c^1, and also how the machine is eventually stopped when the bobbins are full.

Figs. 12, 13, and 14, show an enlarged plan and elevations of the apparatus for this purpose; m^3 is a cast-iron plate fixed to the frame A, and having a projection o^3 at each end. Through the holes in o^3 passes a cylindrical rod n^3, which is omitted in Fig. 14. Upon this slide two forks fixed to a species of rack q^3, with teeth at top and bottom, which are unequally spaced, and the number of which depends upon the number of layers upon the bobbins, generally 20 or 22. Upon the casting m^3 are fixed the two ratchets w^3 x^3, the last of which is kept against the rack by means of the balance weight y^3. The rod b^3 puts these ratchets alternately out of gear by a small up-and-down motion, which it receives by means of the movable frame p at the end of each motion giving a slight push to the rings h^3 attached to the rod. As the ratchets are put out of gear, the rack is at liberty to move slightly forward in the direction of the arrow, until it is stopped by a tooth coming in contact with the other ratchet. The manner in which the rack receives this motion is as follows:—

r^3 is a stud fixed upon the same, working in a slot at the end of the bell-crank t^1, which we have already noticed. Upon the rod n^3 is fixed, by means of the bush z^3, the slotted piece s^1; this is attached to the lever t^1 by the fork u u, capable of sliding up and down the same, and of being fixed in any part of the slot s^1 by the set screw a^4 Motion is imparted to the rod n^3 by a cord b^4 and the weight f^4, pulling at it whenever one of the ratchets are lifted. As it is in connection with the rack through the lever t^1, it is obvious that according to the position of the fork u in the slot s^1, the motion of this latter, and also of the rod n^3, will be more or less proportioned to that of the rack q^3. This motion is finally imparted to the conical drum by means of the rod q^1, which is forked at o^2 to fit on the neck of the drum, and which is fixed by means of a set screw into the bush r^1 on the slotted piece s^1. The pulley d^4, over which the cord passes to the weight f^4, is attached to the bell crank g^4; this, when the machine is at work, is kept in an upright position by the rod c^4 being hooked on to the part o^3 of n^3, but when the rod n^3 has arrived at the end of its motion, that is, when the bobbins are full, the bush p^3 comes in contact with the end of this rod, lifts it out, and as the bell-crank has now nothing to keep it upright, it is pulled down by the weight f^4. In doing so, the upper end strikes against the tappet i^4 fixed to the rod k^4, which is thus pushed forward, and puts the machine out of gear by transferring the driving belt from the fast pulley *a* to the loose one *b*. This can of course also be effected at any time while the bobbins are filling, by moving the rod k^4 with the hand. A bobbin and fly-frame, such as just described, is capable of producing 6 pounds of thread (1,000 yards to the pound) per spindle in twelve hours. One person can take charge of two frames, join the broken threads, and take away the full bobbins.

123. The bobbins from the bobbin and fly-frame are now put in the *fine roving or bobbin and fly-frames*, the construction of which is so similar to the machine just described, that a further description is unnecessary.

124. The actual *spinning process* is performed either by what are termed the *water-twist* and *throstle*, or by the *mule* machines.

The *water-twist* is now almost entirely superseded by the *throstle-frame*, which is shown in Figs. 15, 16, Plate X. Here the twisting and winding-up occurs simultaneously. The general construction is very similar to the roving or bobbin and fly-frames. The bobbins from the fine roving frame are placed at *a*, the thread from which passes through the drawing rollers *b c d*, and then to the spindles. Here it is twisted and wound upon the bobbins as already described, with this difference, that the bobbins *h* have no rotatory motion imparted to them, but simply rest upon the trays *l*, and are dragged round by the thread as it twists itself round them, as the fork or fly *i k* revolves much more rapidly than the thread is given out from the drawing rollers. The up-and-down motion of the bobbins is effected in the following manner:—

The lever *o* turns at its centre on a spindle fixed to the framework. At its ends it is connected by the rods *n n* to the trays *l l*, upon which the bobbins rest, and receives an oscillating motion from the heart-shaped cam *u* through the lever *s t* and rod *q*. The cam depresses the one side of the lever, which is pulled down on the other side by the weight *r* as the cam recedes. By this means the trays, and therefore also the bobbins, are moved up and down. The spindles receive their motion from the drum *y* and the belts *x x* passing round the small pulleys *w w*. Fig. 10 shows the principal gearing, where the wheel k^1, fixed upon the same shaft as the heart-shaped cam, receives its motion from the worm *i*. The pulley *z* makes 600 revolutions per minute, and as the small pulleys on the bobbin spindle are one-sixth of the diameter of the drum *y*, they revolve 3600 times per minute.

125. *The mule* differs in its operation from the *throstle*, inasmuch as in the former the twisting and winding of the thread do not occur simultaneously as in the latter, a length of about 5 feet being first spun and then wound upon the spindles themselves, instead of upon reels or bobbins as formerly. Fig. 1, Plate XI., represents a sectional elevation of the mule. It will be seen that this machine consists of two principal parts,—the fixed frame *z* carrying the bobbins a^1 and the drawing rollers *e f g*, while the movable carriage contains the spindles by which the thread is spun, and upon which it is afterwards wound up.

The thread from the bobbins passes through the loops b^1 b^1, and over the bearers *w n* to the drawing rollers. The bearer *w* has small wire loops through which the thread passes, and receives a slight backward and forward motion through the contrivance as shown in Fig. 3, Plate XI. The drawing rollers have the same construction as already described.

The carriage consists of a cast-iron frame at each end, as shown in the end view Fig. 2. To these are fixed the wooden bearers b^2 c^2 d^2 e^2, which are further strengthened by intermediate iron frames f^2. To it are attached the iron wheels i^2, whose rims have a V-groove, and which run upon corresponding rails, b^2, fixed to the floor of the factory.

The spindles stand in a frame consisting of the wooden parts x^2 x^2, connected by means of the iron pillar y^2, which is placed in a slightly slanting direction. The bottom portion has the small brass cups for the lower end of the steel spindles to rest in; these have small grooved pulleys g^2 fixed upon them, by which they receive their motion from the drum c^3. The machinery by which this drum, of which there is one to every 24 spindles, is made to revolve, and by which the carriage is also made to move backward and forward, is placed in the centre of the mule, and is shown in Fig. 4, Plate XI., in elevation, and in Fig. 5 in plan. The fly wheel c^1 receives its motion from the main shaft of the factory, through the fast and loose pulleys f^1 g^1; round the circumference of c^1 is a groove to receive the crossed cord d^1 by which it drives the pulley e^3, and therefore also h^3. At the other end of the fixed frame opposite h^3 is another pulley i^3, and on each side of the frame we observe in the movable carriage the grooved guide rollers o^3 (Fig. 2), standing at the same angle as the drums c^3, which have also two grooves at the top. One endless cord connects all these drums together in the following manner.

From the top of h^3 it passes, as shown by the dotted line, first round the bottom groove of k^3, and off to the left-hand side of the carriage, where it is wrapped round all the drums c^3, comes back over the guide roller o^3 to the top groove of k^3, and then over i^3, which is a stretching pulley, back to h^3. The drums c^3 on the right-hand side are driven from the centre groove of k^3.

When the carriage is at the commencement of its motion, the tops of the spindles are just in front of the drawing rollers, and the frame d^3 is in contact with q^3. The drawing rollers now commence to turn and deliver the thread which is partially spun by the spindles, the carriage at the same time receding with a slightly greater velocity than that of the front drawing rollers, so that the thread becomes still more stretched. When the carriage has completed its motion backwards, the drawing rollers stop, but the spindles still continue to revolve to complete the spinning. The carriage then moves in again, during which time the thread is wound upon the spindles.

We have now to describe how the backward and forward motion is effected, and the drawing rollers stopped and started.

The axle p^1 of the wheel m^1 which drives the drawing rollers, has a small pinion q^1 in contact with the wheel s^3. The shaft of this wheel is carried by the brackets u^3 and v^3; the former is fixed to the framework, but the latter can turn upon the spindle w^3 in such a manner that s^3 can come out of gearing with q^1. This bracket is connected with the lever x^3, having the balance-weight y^3 which tends to pull s^3 out of gear. At the end of the shaft of s^3 there is fixed a pulley z^3, and a corresponding one q^3 is at the other end of the frame. A rope r^2 runs over these two, the ends of the same being connected with the rod s^2 fixed to the carriage. Upon the fly-wheel shaft c^1 is fixed a bevil-wheel h^1, in gear with a larger one i^1. This is upon the slanting spindle k^1, at the lower end of which is another bevil-wheel l^1, in gear with m^1. The bevil-wheel i must be capable of being put out of gear with h^1, for which reason the bracket n^1 can be moved round by the vertical spindle o^1.

We now see that when the fly-wheel revolves in the direction of the arrow, the carriage will be drawn by means of the rope r^2 away from the drawing rollers until it meets the stop t^2. At the same moment that this takes place, the carriage lifts the bent lever w^3, the other end of which, d^3, is consequently depressed, and frees the stud f^3. This stud is fixed in the slot of the rod g^3, which rod is connected with the bent lever j^2 turning on l^3, the short end of which keeps down the lever x^3. As soon, therefore, as the stud f^3 is freed, the rod g^3 moves forward, being made to do so by the weight y^3 pressing up the lever j^2. The motion of the lever x^3 in consequence puts the wheel s^3 out of gear, thus stopping the backward motion of the carriage, while the forward motion of the rod g^3 turns the spindle o^1, and, by pulling the bevil-wheel i^1 out of gear, stops the motion of the drawing rollers. After both the carriage and the drawing rollers are stopped, the spindles have still to be turned for a certain time to complete the spinning, when the machine is stopped altogether, and the carriage is moved back, and the spun cotton wound upon the spindles by turning the handle e^1 by which the rope r^2 is turned the reverse way.

As the continuation of the spinning only has to take place for a fixed and definite time after the carriage has stopped, the following contrivance has been adopted to throw the strap r^3 on to the loose pulley at the end of that time. At the end of the fly-wheel shaft is fixed a worm t^3 working into the worm wheel u^2, on the spindle of which is the tappet v^2. This tappet, when the revolution of the wheel u^2 is completed, lifts the lever a^4, which till then was in contact with the stop b^4. This lever, being fixed to the belt lever c^4 (to which is also attached the weight k, trying to pull the same over), when lifted away from b^4 allows the belt lever, and consequently also the belt, to be carried over to the loose pulley. In moving back the carriage and winding up the spun cotton, the spinner depresses the lever h^5 g^5, so as to wind the cotton upon the spindles in the manner shown.

When the carriage has again arrived in front of the drawing rollers, it is necessary that the different parts should be put in gear again; this is effected by the carriage pushing against the lever j^2 and q^5, the former putting the wheel i^1 in gear, and by means of the lever x^2 also and the wheel s^3; the latter, q^5 throwing the strap on to the fast pulley again. This completes the series of machines employed in the operation of spinning. We have now but to mention the amount of motive power required for each of the different machines.

	H P
One batting machine, with two batting arms and one blower	3
One spreader, with one batting arm and one blower	2
One breaker card of 18 ins. width	0·13
One finishing card of 3 feet width	0·22
One drawing frame	0·041
One bobbin and fly frame	0·0085
One can roving frame	0·0238
One throstle frame	0·0095
One self-acting mule	0·00228

SECTION III.

126. *Machinery for the Manufacture of Paper.*—In the manufacture of paper, the principal features consist:

Firstly, in the comminution in water of certain fibrous materials, and the formation of a perfectly homogeneous pulp from the same.

Secondly, in the spreading of the pulp in layers of uniform thickness over a wire gauze, withdrawing the water, subjecting the remaining layer of pulp to pressure, and drying the same, thus obtaining a coherent material, consisting of exceedingly small fibres interlaced to a certain degree.

The principal material from which paper is manufactured is linen rags, but in these days of cheap press, where paper is in such enormous demand, other fibrous materials, such as straw, hemp, wood, cotton &c., are also largely used, and in many cases a certain per-centage of paper already used is mixed with the other material.

127. The processes of manufacture may be classified under the following heads:—*a*, sorting of the rags; *b*, cutting them up and dusting them; *c*, washing them; *d*, reducing them to *half-stuff*; *e*, bleaching the *half-stuff*; *f*, the working up of the half-stuff to a fine pulp; *g*, sizing the same; *h*, spreading the pulp to form paper; *i*, compressing the same; *j*, drying the same. The *sorting* is performed by manual labour, 40 to 60 women being necessary to sort from 30 to 45 cwt. in a working day of 12 hours.

The rags are next *cut up* by means of a machine, consisting of a solid cast-iron cylinder, 18 inches in diameter and of the same width, upon which two spiral blades are fixed. This drum revolves at the rate of from 100 to 150 revolutions per minute, and consumes from 3—4 H.P.

After being cut up, the rags are then *dusted* in a conical perforated drum from 5 to 6 feet long, 3 feet 6 inches in diameter at one end, and 2 feet 6 inches at the other, with holes about $\frac{1}{4}$ inch square, inside which drum a wooden shaft with blunt wooden spikes is made to revolve at the rate of from 28 to 36 revolutions per minute.

128. The process of washing is performed in the *half-stuff engine*, which consists, as shown on Plate XII., Figs. 2 to 4, of an oblong vat *a* of cast-iron, $\frac{3}{4}$ inch thick, rounded at the ends, 9 to 11 feet long by 4 feet 6 inches to 6 feet wide, and 2 feet to 2 feet 6 inches deep. In this is fixed, slightly out of the centre, the partition *b*, leaving an opening of about 2 feet from each end.

In the wider side of the partition is placed the cylinder *c*, fixed upon the shaft *d*, which passes across the vat. This cylinder is generally of wood, from 2 feet to 2 feet 6 inches long, and 2 feet to 2 feet 3 inches in diameter; into grooves round its circumference, running parallel with the axis, are fixed a number of steel blades *e*, with a chisel edge. These blades, of which there are about 40 in the *half-stuff* engine, are fixed three in each groove, by means of an iron ring *f*, shrunk on to projections at each end of the blades. Sometimes the three blades in each groove are formed in one, as at Fig. 6. The wooden cylinder is fixed upon the shaft by means of the iron ring *g* and wedges *h*.

In the bottom of the vat, directly under the cylinder, is formed a recess into which are fixed similar cutters to those on the cylinder, but placed slightly at an angle to the axis of the latter. The distance between the cutters on the cylinder and those in the bottom of the vat is variable, by reason of the cylinder-shaft being fixed upon two levers *i i*, having their fulcra in the standards *j j*, and in gear at their other end, with the endless screws *k k* fixed to the standards *l l*. By means of these the shaft *d* is capable of being slightly raised or lowered. The part *m* of the bottom of the vat is formed so as to slope up to the cylinder, having a groove *n* into which any grit is allowed to accumulate; the portion *o* behind the cylinder is formed according to its curve, and then suddenly slopes downward, by which arrangement a constant circulation is caused to take place in the stuff, as indicated by the arrows. The cylinder is covered in with the hood *p*, to prevent the same by its rapid motion from throwing out the stuff. On one side of this hood is fixed the sieve of wire cloth *q*, which can be covered with the lid *r*. This sieve allows the foul water from the stuff that is thrown against it to pass through, which is carried away through the aperture *s*; this is however only done when the washing is going on.

The vat is first filled with the cut rags, to the amount of about 50lbs., and water; the cylinder, being now at its greatest distance from the bottom, revolves at the rate of from 100 to 150 revolutions per minute. After a certain time the cylinder is gradually lowered, when the comminution of the rags takes place. During this time the foul water is allowed to pass through the sieve, while fresh water is continually being supplied. When the cylinder is at its lowest position no more water is allowed to pass away, and the comminution is continued until the *half-stuff* has acquired the proper degree of fineness. This is the case after working for about two hours, when the *half stuff* is allowed to pass out of the vat to the *whole-stuff* or *pulp* engine, to facilitate which the half-stuff engine is generally placed above the pulp engine. The half-stuff engine requires, when washing the rags, from 1½ to 2 H.P., but when the cylinder is at its lowest position, from 4 to 6 H.P.

The half-stuff generally undergoes a bleaching process by means of chlorine.

129. The *pulp machine* differs in no respect from the *half-stuff engine,* excepting in the number of cutters upon the cylinder, which amount generally to 60 and are sharper. The cylinder also revolves with a greater velocity, making from 180 to 200 revolutions per minute. 50lbs. of dry half-stuff constitute a charge for the pulp engine, and this will take about two hours to reduce. If worked too long, the paper prepared from the pulp will have too little tenacity. The pulp engine requires 4 H.P.

For many kinds of paper the *sizing* takes place in the pulp engine, after the charge has been reduced to pulp. There are different compositions which are used for this purpose, the most general being a mixture of about 1½ pound of finely ground alum with 1 pound of wax and ¼ pound o potash, the two latter ingredients being first boiled together in water, with the addition of a small quantity of fresh slaked lime. After the pulp is thus prepared it is run off into the stuff chest, ready for formation into paper. All paper used formerly to be formed by hand, the operation consisting in dipping square frames, over which fine wire cloth is spread, into the pulp, withdrawing a certain quantity, from which the water is allowed to strain off, the workman giving the frame a horizontally vibratory motion the while. The sheet thus obtained is turned on to a felt, and a number of them being piled one on top of the other, a felt always intervening, they are put into a press and afterwards hung upon a line to dry. At the present time, however, only drawing paper is manufactured in this manner, almost all other kinds of paper being produced by the *automaton endless paper machine.* This machine was first invented and patented by a Frenchman of the name of *Robert*, but was afterwards considerably improved in England by *Donkin.* There are several varieties of this machine, but the feature which is common to all is an endless wire cloth strained over rollers and slowly moving in one direction, on to which the pulp is allowed to run evenly at one end.

130. Fig. 1, Plate XII., represents the most general arrangement of this highly ingenious machine. From want of room a portion of the same, continued from the section line x x, is placed below.

The endless paper machine properly consists of a combination of five different apparatus, each performing a separate process, viz.: 1, the stuff chest, in which the pulp is stirred up and strained; 2, the before-mentioned endless wire cloth, upon which the paper is first formed; 3, a series of compressing rollers; 4, a series of drying and smoothing rollers; and 5, an apparatus for winding up the paper as it is delivered from the rollers.

The figure represents a longitudinal section through the machine. The pulp first enters the compartment A, where it is mixed with a certain quantity of water from the cock A^1, and is stirred up by means of the agitator a. There are four small sluices b, through which the pulp passes into the strainer B, consisting of four compartments. This strainer is hinged at b^1, and receives through three arms b^2 and the revolving cam b^3 a vibratory motion, by which it allows the pulp to pass through, and retains the knots and other impurities. In the compartment A^2 the pulp is again stirred up by means of the agitator a^1, and thence passes through the long narrow sluice C over c^1 and under c^2 on to D, where any sediment is carried off. From D the pulp passes in a perfectly even layer on to the endless wire cloth E. This is of fine brass wire, from 4 feet 6 inches to 5 feet wide, and is carried by 44 small hollow copper rollers *e* so as to form a perfectly level surface. Near the stuff chest these rollers are placed quite close together, but further on small spaces are left between them, to allow the water to pass away more freely. The cloth is afterwards carried by several larger rollers at greater distances, passes over the exhausting apparatus I and then round the drum K, and over the straining rollers f^2 back round the roller d^2.

This endless cloth has a twofold motion imparted to it: Firstly, the continuous one as indicated by the arrows; and secondly, a quick horizontal vibratory one in a direction at right angles to the other motion. The object of this latter motion is to cause the pulp to lie evenly on the cloth, as also to interlace the fibres as much as possible, in order to increase the tenacity of the paper. It is imparted to the whole of the frame carrying the endless cloth rollers &c. by means of a small crank and connecting rod attached to one of the oscillating supports *ff*, of which there are five pair carrying the frame.

131. In order to regulate the width of the paper, an endless band or *deckle* *g*, is placed on either side of the endless cloth, moving along with the same. These are about ½ inch thick and formed of layers of tape fixed together by a solution of india rubber. The deckle straps pass round the guide roller g^2 and are strained by means of other rollers g^1; they are cleansed from any particles of pulp that may adhere to them by means of a continuous stream of water from the cock g^3, which water is caught up by the trough beneath. The water passing through the wire cloth E is also caught up underneath in the trough H, and is carried back to the vat. From the small

compartment h^2 a broad stream of water is made to play upon the drum d^2 to clear it from pulp, which water is supplied from the pipe h^3. Other cocks h^4 are provided at the other end of the wire cloth, to cleanse the same from adhering particles of pulp.

To cause the paper pulp to be freed more rapidly from the remaining water, the cloth passes over a trough I, which is connected with an air-pump for exhausting the air partially out of the same, thus creating a vacuum of about one-sixth or one-eighth less pressure than that of the atmosphere. This pressure condenses the paper pulp and extracts the greater quantity of water from it. The wire cloth is supported, while passing over the trough, by four small rollers, and the water passes from the latter through the syphon bend *i*.

After having passed over the trough, the pulp is further compressed, first by the small roller W, which is also covered with wire cloth, and afterwards by the so-called *first press*, consisting of rollers K and K^1, round the former of which, as already mentioned, the wire cloth passes. These drums are of copper and are covered with woollen felt.

The drum K^1 is pressed upon K by means of the lever and weight k^1, the bearings of its spindle being enabled to slide upon an inclined plane. It is also furnished with a scraper and a stream of water from the cock k^2 for cleansing purposes.

After passing between the drums K K^1, the paper pulp, which has now already acquired some consistency, leaves the endless wire cloth E and passes over to the endless felt *m*, strained by the eight rollers M, which carries the paper to the *second press*, that is, between the two smooth cast-iron rollers N N^1. The upper one is in immediate contact with the paper, while the lower one has the felt intervening. The degree of pressure which these rollers exert is regulated by means of the set screw n^1; they are also provided with scrapers for clearing them of any particles of paper.

The felt is strained in its length by the roller M^1 adjustable on the rack m^1, and in its width by the nippers *qq*, which are furnished with two small revolving cones between which the felt has to pass. From the felt *m* the paper passes over the rollers O O^1 (as indicated by the dotted line) to the *third press* consisting of the cast-iron rollers P P^1, similar in all respects to the *second press.* By the arrangement of rollers it will be seen that the side of the paper, which, at the *second press* was in contact with the felt, in the *third press* is in contact with the smooth roller P^1. Hence the paper passes over the rollers Q Q to the drying apparatus, consisting of a series of large hollow cast-iron drums R T V, round the greater part of the circumference of which the paper is in close contact, being covered by the felts. These drums are continuously heated by means of steam, passing from the pipe S into the interior of the same through the small pipe S^1 and out again through S^2. At the second drum T the paper is subjected to a slight pressure from the small drum T^1; this is termed the *first dry press*, and the paper here receives a smoothed surface on one side, the other side being served in the same manner by the *second dry press* V^1. The reason of this being placed rather above the other steam cylinders lies in the necessity for keeping the paper in a slightly moistened state while it is being hot-pressed, which is thus effected by the vapour rising from the cylinder T.

132. Upon leaving the second *dry* or *hot press* V, the preparation of the paper is completed, and it is now wound upon the hexagonal expanding reels Y and Y^1, which, when one has received the desired quantity (generally consisting of about 60 layers), are swung round with the lever *y*, so as to change their relative positions. The rotatory motion imparted to the revolving reel must be a differential one, that is, as the quantity of paper delivered from the press is always constant, the reel must be made to revolve gradually slower as its diameter is increased by every successive layer

of paper. The circumference of the reel may be increased or decreased at pleasure to suit the size of sheet required.

133. The speed of the drums, rollers, &c. must be adjusted to a very great nicety, so that the paper shall everywhere be passed onward with a constant linear velocity. The motion is generally communicated in the first place to the *second wet press* N, whence all the other drums, &c. receive motion, either by wheels or by straps. The cylinder N generally makes about 12 or 13 revolutions per minute, and as its circumference is about 30 inches, the quantity of paper delivered per minute is from 30 to 32½ feet.

The requisite power for working this machine amounts to about 4 to 5 H. P. As the speed is variable according to the description of paper to be manufactured, it is generally advisable to have a separate motor for this machine. It requires, per minute, four cubic feet of water, and three or four attendants. The requisite number of stuff engines for such a machine are, of half stuff engines from four to five, and of pulp engines from six to eight. For these, four to five attendants are necessary. A steam boiler of about 150 square feet heating surface will suffice for supplying the steam cylinders, for boiling the rags and the size. This will give a consumption of about 1lb. of coal per production of 1lb. of paper. (The requisite amount for heating the various rooms would be additional.)

134. Besides this description of paper machine, there is also another much used, which differs essentially from it with respect to the first formation of the paper. Instead of the pulp being delivered on to a horizontal wire cloth, there is a large cylinder covered with the same description of wire cloth, which is half immersed in the pulp vat. The ends of this cylinder being closed, and the percolating water being extracted through the hollow spindle, there results a pressure upon the circumference of the cylinder due to the difference of level of the water inside and outside of it, thus causing the adhesion of a layer of pulp on to the

cylinder as it revolves, which is taken off and undergoes compressing and drying processes, the same as in the machine just described.

SECTION IV.

135. *Machinery for the Manufacture of Conical Bullets.*—In the class of machinery now under consideration, we have a practical illustration of the application of cams to effect certain distinct and consecutive operations, performed in one and the same machine upon the material to be manufactured, which operations, without the aid of these automaton contrivances, would be almost impracticable.

The *Conical Bullet machine*, the invention of Mr. J. Anderson of the Royal Arsenal, Woolwich, has for its object the rapid and accurate manufacture of that peculiar projectile called the *conical or Minié bullet*, which has now entirely superseded the spherical bullet formerly used for rifled small arms. The principal advantages of this method of manufacture over the old method of casting the bullets are: firstly, the much greater rapidity with which they are made, and, secondly, the more perfectly homogeneous nature of the projectile, which latter quality is very essential, to accuracy of firing.

136. Plates XIII. and XIV. represent the elevation and plan of this very ingenious machine. Before however going into a detailed description of its action, we must briefly describe the manner in which the lead has first to be prepared in the form of a continuous cylindrical filament before being subjected to further operations. The lead is run in a molten state into a massive iron cylinder whose exterior diameter measures 2 feet 6 inches, the interior diameter being 7½ inches, by a depth of 2 feet; the quantity for one charge being about 4 cwt. When the lead is in a semi-solidified state, a ram or plunger accurately fitting the bore of the cylinder, and provided with a circular opening through its whole length, is made to descend and press

upon the lead by means of hydraulic pressure. The lead is thereby caused to ascend through the circular die at the bottom of the plunger, and comes out at the top of the same in the form of a continuous cylindrical filament as already stated. It is then wound upon large iron reels, and is conveyed to the machine now to be described.

137. In Plate XIII., Fig. 1 represents the side view of the bullet machine, and Fig. 2 the end view, while in Plate XIV. is represented a sectional plan of the same. The machine contains four *working points*, all of precisely the same construction, one and the same cam and die shaft serving for each two upon one side of the machine.

The mechanical process consists in: 1, unwinding a portion of the lead filament from the reels; 2, detaching a small portion of the same; 3, forcing the detached portion into the die or mould, whereby it is made to acquire the requisite shape; 4, freeing the bullet from superfluous portions; and 5, ejecting the bullet from the die.

138. Upon four projecting axles *a*, fixed to the cast-iron framework A, are placed the four reels B B, containing the lead filament as already described. These reels are furnished upon one side with a toothed circumference into which act the pinions *b*, driven by means of the drums C on the shaft overhead, and the fast and loose pulleys D D, upon the same spindle E as the pinions. By these pinions the reels are caused to revolve slowly in the direction of the arrows, so as to unwind the lead filament. This is then conducted through the loops F, and between the guide rollers G, to the nipping rollers H. The lead having to be brought forward to a certain length, and then to remain stationary while the portion is being detached, it follows that the nipping rollers cannot have a continuous rotatory motion, but must be made to revolve periodically. This is effected by a ratchet wheel *h* being fixed upon the axle of the lower one of each pair of nipping rollers, which ratchet wheel is acted upon by a series of pauls h^1, receiving their

reciprocating motion through the connecting rods h^2 from the loop I of the reciprocating die spindle J. The motion of this, as also of the cam shaft K, is communicated in the following manner.

The drums L L upon the shafting overhead communicate motion to the fast and loose drums M M^1 upon the same shafts N, with which are the pinions O working into the spur wheels P. These shafts are also provided with fly wheels Q.

The shafts of the spur wheels are carried at *p p* by the framing, and at p^1 are formed excentrically, as seen more clearly at the enlarged section Fig. 2, Plate XIV. This excentric imparts a reciprocating motion to the die spindle J by means of the sliding block R working in the loop J^1. Upon the ends of the shafts *p* are fixed the bevil pinions S working into the mitre bevils S^1 fixed upon the cam shafts K, which thus receive their continuous rotatory motion.

139. We will now return to the lead filament which we left between the nipping or feed rollers H. After leaving these it passes through the slightly tapered hole t^1 (see Fig. 2, Plate XIV.) in the lever T. This hole is formed in a bush of steel, and on the side furthest from the feed rollers it is just the diameter of the lead, the edge being made very sharp. On the front of the lever T, just opposite the hole just mentioned, is fixed an adjustable stop, the object of which is to limit the exact length of lead which is thrust through the hole, for this is the piece which has now to be severed off. To effect this, a pair of nippers U, of the form as shown in Fig. 3, Plate XIV, are placed close against the lever T. Each of these nippers is fixed to a spindle *u* held by the framing A, on the other side of which one of these spindles is attached to the lever V, the other being made to work with it by means of the toothed segments u^1. When the lead is thrust through the hole in the lever T, the nippers U are made to close upon and hold fast the protruding piece, which is effected by the lever V being raised

by the cam v upon the shaft, thus turning each of the nippers slightly on their axes, and causing them to close at the top.

When the nippers have thus laid firm hold of the lead, the lever T is raised by means of the cam t upon the cam shaft K, and the leaden filament held in the hole t^1 of this lever, being raised with it while the protruding portion between the nippers remains stationary, the latter is severed from the former by a kind of shearing action.

As soon as this has taken place, the lever V drops suddenly, owing to the receding of the cam V, and the nippers U open at the top and close at the bottom, being then in the position as shown in Fig. 3, Plate XIV. The small leaden cylinder being now let free, falls by virtue of its gravity, until it is caught up at the bottom of the nippers, which of course must have closed before the lead arrives there. This brings the leaden cylinder exactly in front of the female die W, into which it is now forced by the male die W^1, fixed to the end of the die spindle J, being brought forward by the action of the excentric p^1 formed upon the shaft p. To allow means for any superfluous lead to escape from the die after it has been forced in, there is a male space left between the two dies, and between them this superfluous lead forms a thin rim, still connected with the bullet. The bullet now formed has to be ejected from the die, and this is done in the following manner.

140. The bush w into which the female die is fixed is hollow, so as to receive the small spindle w^1, to the end of which is attached a small rod w^2 passing through to the back of the female die, and forming the bottom of the same. This spindle is moveable backwards and forwards in the bush w. When right back, the end of the rod w^2 forms the bottom of the die, and when right forward it is in the position as shown in Fig. 2, Plate XIV., having by this movement ejected the bullet from the die. The spindle w^1 is actuated by

means of the lever X, which receives its motion from the cam X^1 upon the cam shaft, being furnished at the end with a small roller, as are in fact all the other cam levers. The action of the cam X^1 is very sudden, so that the spindle jerks out the bullet, and is brought back into its original position almost instantaneously. There is an adjustable stop x fixed at the back of the spindle, against which this latter is brought by the lever X. It must here be mentioned that there is a minute perforation through the whole length of the spindle w^1 and the rod w^2, which serves to allow egress to the air contained in the female die when the lead is being pressed in; this perforation is however too small to allow any lead to pass up it.

141. We have before mentioned that the bullet when formed in the die has, outside of the latter, a thin rim of lead attached to it. In order to remove this, a steel plate Y, containing a circular hole the exact size of the bullet, is raised in front of the female die after the male die has receded, and through the hole in this plate the bullet has to pass when ejected, thus leaving the thin rim of lead behind, which falls to the ground. When the plate Y drops again, a fixed knife-edge passes over the hole, and clears it of any particles of lead that may be adhering to it. This is essential, as upon the clean nature of the hole depends the accuracy of the calibre of the bullet. This template receives its motion from the lever Y^1 and cam Y^2, and is brought down by the weight y^1 and spring y^2; it slides between guides, adjustable by means of the set screws y^3. The bush w, containing the female die, is also adjustable with great accuracy by means of the eight set screws w^3.

When the male die has receded to its most backward position, as shown in the sectional plan Fig. 1, Plate XIV., a small steel blade covered with flannel saturated with oil, is brought rapidly forward by means of the spindle Z, actuated by the cam Z^1, and passes over the end of the die, thus oiling the same. This is necessary both for preventing the bullet from adhering to the part of the male die projecting into it, and for keeping the die cool.

It will be observed that the cam *v* working the lever V attached to the nippers U, after rising suddenly and continuing concentric for some time, has then another small rise upon it. The reason of this is, that the nippers must open at the bottom to allow the template to rise in front of the die, but they cannot quite close at the top until the lead has been brought quite forward, which does not take place until shortly afterwards.

The bullets, after being shot out of the die, pass down the trough *c* into the box *d*, where they are collected.

Each of the dies turns out 30 finished bullets per minute, which for the whole machine would give for a working day of 12 hours, the enormous number of 86,400.

SECTION V.

142. *Machinery for the Conversion of Timber.*—The operations of machines employed in this branch of manufacture are of a threefold description, namely, *sawing*, *planing*, and *moulding*. By sawing machinery we effect the subdivision, into two or more parts, of a piece of timber, either by an intermittent process, as with the *timber-frame* and *deal-frame*; or by a continuous action of the saw, as with the *circular saw*, the *cylindrical saw* and the *ribband saw*. By planing machinery, the rough surfaces of the timber, resulting from the operation of the saw, are rendered perfectly level and smooth; the different methods of performing this are, firstly, by means of cutters fixed upon a horizontal disc revolving with considerable velocity round a vertical axis, the cutters operating upon a surface parallel with the disc; secondly, that of long cutting edges fixed to spindles in a direction parallel to their axis, and revolving with great rapidity with them, the cutting edges describing a cylinder, and operating upon a surface parallel with their axis of rotation; thirdly, the

action resembling that of a carpenter's plane, the cutting edge being fixed, and the surface to be planed passed over the same.

In *moulding machinery* the cutting edges of the tool have the exact form of the moulding to be produced. Four or more of these are fixed upon a spindle in a direction parallel to its axis of rotation, as shown in Fig. 8, Plate XV., similar to the second described method of planing, and while revolving with great velocity, these templates cut into and scoop out the piece of wood under operation so as to give it the particular form of moulding required, the wood being fed forward as it is operated upon. According to the nature of the moulding, its formation requires one or more operations of such tools; if the latter is the case, the first operation gives the wood a preparatory outline of the required moulding, so as to render the work easier for the finishing cutters, and thus preserve the keenness of their edge and accuracy of their action necessary for turning out good work.

143. The *timber frame* is the sawing machine usually employed for cutting up large balks of timber into lengths of planking; this is performed by means of a series of saws of a coarse pitch fixed at certain distances from, and parallel to, one another in a strong iron frame. To this frame is imparted a reciprocating motion, during the up-stroke of which the timber is fed forward, while during the down stroke, by virtue of the saws being placed at a slight incline, namely, rather more forward at the top than at the bottom, they cut into the wood to an extent due to this inclination.

144. In Figs. 1, 2, and 3, Plate XV., are shown side view, front view, and plan of such a *timber-frame,* as constructed by Mr. Horne, engineer, of London. It is capable of cutting up balks 3 feet 6 inches large, and is furnished for this purpose with 13 saw blades at 3 inches distance, that being the usual thickness of large planks. *a a* is a strong

cast iron framework, bolted down firmly to the foundation. Below the floor-line *b* of the saw-mills is fixed the driving-shaft *c*, carried by the plummer-blocks fixed to the framework and foundations; upon this shaft are the fast and loose drum *d d'*, which receive their motion from the main shafting driven by the engine. In addition, the shaft carries the fly-wheel *e* and crank-plate *f*; the latter imparts a reciprocating motion to the saw-frame *g* by means of the connecting-rod *h*. The saw-frame consists, firstly, of the transverse bearers *g* at top and bottom, which are formed of two wrought iron cheeks fixed together, with such a distance intervening that the loops to which the saw-blades are rivetted can pass between them and receive the cotters *i* by which the blades are strained, on the outside. These two transverse bearers are connected together by means of the upright tubes *j*, and the lower one is furnished with the cheeks *k* to which the connecting-rod is fastened. At their ends these bearers are so formed as to slide up and down the guide bars *l* fixed to the frame *a*.

145. Upon the longitudinal beams *m*, to which the framework is fixed, the rails *m'* are laid, to an extent equal to twice the length of the timber-carriage which runs upon them. This carriage, which is of a length equal to that of the timber to be cut up, consists of the cast-iron frame *n*, carried by the small grooved rollers which run upon the rails *m'*, and furnished at *o* with a toothed rack into which work, on one side, the pinion *p*, on the other the pinion *q*. The two sides *n* of the carriage are connected by means of the adjustable transverse pieces *r r'* of which *r* may be fixed at any place upon the side frames, while *r'* is adjustable by means of the screw *s*. Both these transverse bearers carry an apparatus by means of which the timber is firmly held in its position. This consists of the casting *t* adjustable transversely to a certain extent upon the spindles *u*; *v v* are right- and left-handed screws, by turning which by means of the handle *w*, the claws *x* are brought closer

together and so nip the timber firmly. These claws hinge upon the moveable pins *y* in the casting *t*. Close in front of the saws the timber is carried upon the bearer *a'* by the roller *b'* and is pressed down upon the same by the roller *c'* and weighted lever *e'*, while behind, it is kept down by the guard *f'*. The timber-carriage receives its forward or *feed* motion through the pinion *p* which is fixed upon the same spindle as the ratchet wheel *p'*, this is actuated by the ratchet *g'* receiving its reciprocating motion by means of the rod *h'* of the excentric *i'* fixed upon the driving-shaft *c*; the rod *h'* is adjustable in the slot of the lever *j'* by which the stroke of the ratchet, and consequently the amount of *feed*, is regulated. The *running back* motion is performed by means of the pinion *q* actuated by the spur wheels *q'* and *k'* which latter receives its motion by the fast and loose pullies *l'*. The timber-carriage is kept down where the pinions *p* and *q* act upon it by the rollers *n'* fixed to the side of the frame *a*.

146. The action of this machine will be readily understood. As before mentioned, the saw-blades are so fixed in the frame that they are slightly canted forward out of the perpendicular; at the same time that the crank *f* and connecting rod raise the saw frame (by which action the teeth of the saw recede from the line of the previous cut), the excentric *i'* depresses the lever *j'* which, by raising the ratchet *g'*, causes the ratchet wheel *p'* to revolve to a certain extent, and thus by means of the pinion *p* to bring the timber carriage forward until the face of the saw-cut is again in contact with the teeth. The crank then causes the saw-frame to descend and cut through a length of timber equal to that represented by the cut of the saw-blades. In order that the blades should move freely in the saw-cuts, the latter are rendered of a greater width than the actual thickness of the saws, by the teeth being *set*, that is, they are alternately bent, one a little on one side and the next on the other side, which is done by means of the blow of a

hammer. The form of the teeth is as shown at fig. 4, they are usually stamped out roughly by a press with dies, and then the edges worked as shown by means of the file and hone. The power required for working these machines naturally varies according to the nature of the timber and the number of saw-blades they carry. On an average they consume 12 HP., and make 110—120 cuts a minute.

147. The *deal frame* differs little in its general construction from timber frame. It is employed to cut up into thin deals the squared balks or planks as produced by the latter, and, its work being of a less heavy nature, the different parts generally obtain a lighter construction, and the teeth of the saws are of a less *pitch* and *set*. Owing to the timber upon which it has to operate being generally cut square previously, the long timber carriage, as in the timber-frame, is dispensed with, and the timber is fed forward by means of *nipping rollers*, pressing either against the sides or top and bottom of the same, which receive their motion from the ratchet-wheel; the timber is in this case supported by a small carriage at each end. These machines require on an average about 5 HP. to work them at 100—150 cuts per minute.

The ratchet-wheel is now frequently replaced by the *silent feed*, a contrivance deriving its name from the absence, in its action, of the noise attendant on that of the ratchet-wheel; it consists of a species of drum, the circumference of which is situated between two plates connected together and to a reciprocating rod, in such a manner that when pulled at by the latter, they nip the circumference of the drum and cause it to move with them, but when pushed back they loose their hold upon the drum and slip back upon its circumference.

148. In some cases, particularly where the timber-frame is not constantly at work, or where the amount of resistance in sawing varies very considerably, the same is worked by means of a separate steam cylinder, of which the piston-rod is connected directly with the saw-frame, the cylinder being

generally fixed to the top of the frame-work, and a connecting rod, fixed to the saw-frame, imparting its motion by means of a crank to a shaft, carrying a fly-wheel and the excentrics for the feed motion. This arrangement is exceedingly advantageous both as regards economy and precision of work; as in the first instance it obviates the expense of long lengths of strong shafting and gearing, as also the loss of power in overcoming the friction and inertia of the same, while in the second instance it enables the workman to regulate the speed of the frame according to the nature of the work on it, which is not in his power where the machine is worked by a drum and belt from the main shafting; in which case its speed is dependent on that of the driving engine. Where, however, there are many such machines working pretty constantly and not situated at very great distances from one another, as is generally the case in a saw-mill, the usual arrangement will be most economical, as there is always a loss of power attendant on the direct acting system, owing to the radiation of heat from the pipes conveying the steam to the different machines; beside which, these cannot work the steam expansively, *i.e.*, economically, to any great degree, if at all.

149. An invention has lately been patented by Mr. Thompson of Woolwich Dockyard, which is of great utility in naval architecture, and consists in the application of the timber frame to form slightly curved and winding surfaces on pieces of timber such as are used for the ribs of ships. These curved surfaces are formed by a contrivance by means of which the saw-blades may be made to travel transversely on the saw-frame as they cut through the timber, the teeth being set to a considerable extent to allow the blades liberty of action. The winding surfaces are made by a peculiar construction of the timber carriage, which enables the timber to be turned on an axis parallel with its length while it is being cut.

150. The *circular saw* consists of a circular disc of steel,

the rim of which is formed into teeth, and which revolves with great rapidity. Its action is continuous, and, on account of its great velocity, very effective; it is, however, on account of the large diameter it would obtain, not generally employed for cutting large timber, though circular saws are made of six feet in diameter.

In figs. 5, 6, 7, plate XV., is represented the elevation, plan, and end view of what is termed a *pulling up bench*, as constructed by Mr. James Horne, the machine deriving its name from a self-feeding apparatus attached to it.

151. The circular saw *a* is in this case three feet diameter, it is fixed upon the spindle *b* and is driven by means of the fast and loose drums *c c'* which receive their motion from the main shafting. The frame-work *d* is formed as a table, through a slot in which the saw projects, the spindle *b* being kept underneath, so that the timber, which has already been cut square, can pass along the table on either side of the saw.

The thickness to which the planks or deals are to be sawn is regulated by the position of the *fence e* which is adjustable by means of the screw *f*, and handle *g* in a direction at right angles to the motion of the timber; the latter is pressed against the fence before coming in contact with the saw by means of the nipping-roller *h*, actuated by the lever *i* and weight *j*. The fence is further adjustable in its vertical position by means of the set screws *k*. To cut what are termed *feather-edged* deals, it is fixed by them so as to form a vertical angle with the circular saw, by which means the deal sawn off by the latter is thicker at one edge than at the other.

152. The self-feeding apparatus is arranged in the following manner:—Upon the end of the spindle *b* are fixed the small pullies *l*, which, by means of the gut *m*, communicate motion to the speed pullies *n*, reducing at the same time the velocity; these again impart their motion to the reversed set of speed pullies *o*, reducing the velocity still more; on

the same spindle with these is fixed a small pinion *p* working into the spur wheel *q*, and on the same spindle with the latter is the pinion *r* working into the spur wheel *s*. The drum *t*, which runs loose upon the same shaft *u* with the latter, receives its rotatory motion, which is evidently a very slow one, by means of the moveable clutch *v*, actuated by the lever *w*, rod *w′* and handle *w″*.

When a piece of timber is placed upon the bench, the clutch is put out of gear, and the rope is run off the drum to the end of the timber where the hook *x* is fixed, as shown; the clutch is then put in gear, and the drum, revolving with the shaft *u*, begins to wind up the rope, and thereby pulls the timber against the circular saw *a*. The velocity of this drum is rendered variable for different kinds of wood by means of the guts being changed from one pair of speed pullies to another.

This machine requires 8 HP. to work it at 350 revolutions of the saw per minute. The velocity of circular saws varies from 250 to 500 revolutions per minute. They are generally fixed singly upon the spindles, but in some particular cases, where pieces of timber have to be moulded into certain shapes, a great saving of time has been effected by causing two or more circular saws, fixed either parallel or at an angle to one another, to act upon a piece of timber at the same time. Circular saws are also often employed as *cross-cut saws*, i.e., to cut through a piece of timber transversely. In such cases the timber is generally stationary, and the saw travels as it cuts through the same.

153. The *cylindrical saw* consists of a cylinder of thin steel, one end of which is provided with teeth, while the other is fixed to a spindle in such a manner that the latter causes it to revolve with a great velocity round its own axis. This description of saw meets with only a limited application, inasmuch as it is only applicable where it is required to produce either cylindrical pieces of wood, or such as are curved to form a portion of a cylinder; as for

instance, the staves of barrels before they are bent. Its action is naturally also continuous.

154. *The ribband saw* is a thin and narrow endless band of steel, one edge of which is serrated, which is stretched over two drums, such a distance apart, that there is a certain length of straight saw band between them. Motion is communicated to one of these drums, which causes the band saw to travel with it by friction. Owing to the small width and thickness of the band it is capable of cutting the wood to almost any curve required, to which fact it owes its present extensive application.

CHAPTER VIII.

OF PUMPS AND PUMPING ENGINES.

SECTION I.

155. *On the General Construction of Pumps.*—The common pump is an object so familiar to every one, and the principle upon which it acts so generally understood, that it is scarcely necessary to describe it here. We shall therefore confine ourselves to the consideration of special applications of the same, and in inquiring into its most advantageous mode of action, describe such modifications of the different parts as obtain in particular cases.

156. Very few pumps except those used for domestic purposes are employed as *lift pumps* only, that is, where the water is merely drawn up a suction pipe of 32 feet length, or under, by means of the atmospheric pressure, and then flows away; those employed for engineering purposes being generally a combination of the *lift pump* and the *force pump;* that is, after the pressure of the atmosphere has been taken advantage of to raise the water about 32 feet, there still remains a certain height to which the water has to be *forced* before it arrives at its destination, which height in many cases, particularly in waterworks, is required to enable the water to flow through some miles of conduit pipe.

157. Besides the distinction of *lift* or *suction* pumps and

force pumps, we classify them according to their mode of action as *single acting* and *double acting*. Pumps of the first class only give motion to the water during either the *up* or the *down* stroke of the pump bucket or piston. They can either be force or lift pumps; if the first, they obtain one suction and one delivery valve, and the piston generally obtains the form termed *plunger*, this is nothing more than a piston rod of a large diameter, without a piston, which in descending, fills out more or less the pump barrel, and thus forces the water through the delivery valve. Where, however, both the valves are placéd immediately over and under the pump barrel, the piston is formed with a valve so as to allow the water to pass through it during its downward stroke. Of this construction is the so-called *air pump* of the common condensing steam engine. The single acting suction pump only has a suction valve, and may also have a *plunger* or a *valve piston*, according to the mode of exit of the water.

The second class, namely, *double acting* pumps, work both during the up and down stroke. The only construction of suction pumps to which this term applies, is that in which the piston is a combination of *plunger* and valve piston, that is, in which the piston rod is made of such a diameter, that in descending it displaces half the volume of water that passes through the valve piston, the other half being raised by the latter when ascending. This arrangement may also be employed as a double acting force pump, in which case a *delivery valve* has to be added. In all other constructions of double acting force pumps, we shall have two suction- and two delivery-valves, each pair being in communication with one side of the piston which is then solid, *i.e.*, without valves.

158. There are two methods of imparting motion to pumps, namely, either by direct action or by crank motion. The former is only unobjectionable in the case of simple suction pumps, or with force pumps, that work with only

a very slight *head*, or pressure on the delivery valve. Where it is used for working *high pressure* pumps having a considerable length of *rising main* (the name given to the vertical pipe from the pumps in a well to the surface of the ground), or in connection with some miles of *conduit pipe*, the disadvantage lies in the suddenness with which the flow of water is checked at the turn of the stroke, particularly if the pump is single-acting, where a considerable time elapses before motion is again communicated; indeed, the direct action would be in most such cases altogether impracticable, were it not for a most important feature which is added to all well-constructed pumps, viz., the *air-vessel*. This is generally a strong cast-iron air-tight chamber, of considerable capacity, into which the water passes directly from the delivery valve of the pump, and from which it flows up the rising main or through the conduit pipes. The pumps, in forcing the water into this air-vessel, compress the air contained in the same until its expansive force is equal to the pressure of a column of water corresponding to the height the same has to be raised.

By means of this elastic medium, the sudden commencement and cessation of the pressure at the beginning and end of the *stroke* is to a certain extent equalised, and that in the following manner: At the beginning of the stroke, were there no air-vessel, the whole column of water in the rising main would have suddenly to be started with a velocity equal to that of the pump bucket, thus absorbing a large amount of power in overcoming the inertia; *with* the air-vessel, however, the elasticity of the air permits of its being compressed somewhat *beyond* that density corresponding to the column of water it balances, until it has *gradually* imparted to the latter its proper velocity. At the end of the stroke, *without* the air-vessel, the motion imparted to the column of water in the pipes would have to be as suddenly stopped as it was started; and if the water be flowing with considerable velocity through a great length of conduit pipe instead of up a rising main, in which case its momentum would not be so suddenly counteracted by gravity, the pump bucket would actually have to pull at the whole mass of water in motion before it began the return stroke, in order to close the delivery valve. Besides the loss of power resulting therefrom, these sudden concussions of the water in the pipes have often the disastrous effect of breaking up the *main,* particularly if there occur any very sharp bends in the pipes. With the air-vessel, however, the *magazine* of pressure, as one may term it, contained in the same, would replace the pressure of the pump at the end of the stroke, and the next stroke would commence without the water in the pipes having come to a stand-still at all.

Thus we see that the effect of the air-vessel as an accumulator and equaliser of power is similar to that of the fly-wheel, considered in section 64.

159. The *crank motion* is by far preferable to the direct action for communicating motion to pumps, inasmuch as its action approaches very near to the theoretically correct motion that should be imparted to the pump bucket, viz., a uniformly accelerated motion from the beginning to the centre of the stroke, and a uniformly retarded motion from the centre to the end of the stroke (such as would be imparted by the cam as shown in Fig. 23, Plate III.). This arrangement is now generally adopted in the construction of all pumps where regularity of flow is required; as, for instance, in water-works' pumping engines, the crank being connected with a fly-wheel, which converts the reciprocating motion of the steam cylinder piston into a uniform rotatory one, and this is again changed, by means of the crank and connecting rod, into the desired increasing and decreasing motion for the pump bucket. A pair of such pumps, coupled together at right angles, working at a considerable speed, and both delivering the water into the same air-vessel, will render the flow of water in the pipes almost perfectly uniform.

160. Another means for equalising the flow of water is in the *stand-pipe*. This is a vertical pipe, open at the upper end, and connected at the bottom with the pumps and mains; it is generally employed where there is a lông length of horizontal suction main, which may occur where circumstances necessitate the erection of the pumps at some distance from the source of water. Its effect is similar to that of the air-vessel, namely, when the stroke of the pumps has set the body of water in the suction main in motion, and the piston arrives at the end of the stroke, this motion finds its vent in the *stand-pipe*, and the momentum expends itself in increasing the *head* of water in the same to a certain extent until the pump bucket has attained its full velocity in the next stroke.

161. Notwithstanding the before-mentioned disadvantages, the *direct acting* or *Cornish* pumping engine is generally employed in mining districts for pumping the water from the deep shafts. These engines are generally *single acting*, and are connected with the pump rods by means of a beam; as the pump barrels are necessarily at a considerable depth down in the well, the unavoidable length and strength of the pump rods make them of a great weight. The power of the steam-engine is therefore actually expended in raising these rods, which then in falling give out this power again in forcing down the pump bucket and thus raising the water. As these engines are generally worked with a considerable degree of expansion, the manner of employing the power is advantageous to their action, as at the beginning of the stroke, when there is consequently a great pressure of steam in the cylinder, the surplus power is expended in imparting momentum to the heavy pump-rods which give it off again at the end of the stroke when the pressure in the cylinder is low.

As the velocity with which the pump-rods descend cannot be varied, the work of the engine is regulated by means of an apparatus termed *cateract*, which causes a certain time to elapse between the end of the down stroke and the beginning of the next up stroke, the duration of which pause can be regulated according to pleasure. This apparatus consists mainly of a small cylinder in which works a weighted piston. When the pump rods descend, they actuate the raising of this piston; and the cylinder then becomes filled with water through a valve, opening inwards. In descending again, the heavy piston forces the water out of the cylinder through a small valve which is set to a certain extent of opening and thus regulates the time that the piston takes in descending. When the piston has arrived at the bottom of the cylinder it causes the moving of a tappet in communication with the steam valves, steam is admitted in the cylinder and the up stroke commences. These Cornish pumping engines have also been employed, until lately, generally for water-works' pumping-engines, in which case the weight of the heavy pump rods is replaced by a *balance bob* of considerable weight. This consists of a receptacle attached to the *plunger* piston, into which is placed a certain amount of ballast, sometimes amounting to more than 40 tons (as in the Wicksteed engine at the East London Waterworks, the engines at the Battersea Waterworks, &c.).

162. The most extensive application that direct acting pumping-engines have yet met with is for draining the *Haarlemmer Mere* of South Holland.* These works consist of three separate engines, each of 35 HP., one of which works eleven pumps of 63 inches diameter with 10 feet stroke; the other two working 8 pumps each, of 73 inches diameter and 10 feet stroke, each engine raising about 66 tons of water per stroke. The pumps are arranged round the steam cylinder and are worked simultaneously from the same by means of beams fixed to a massive cross-head carried by five piston rods. These beams pass through

* See "Dictionary of Terms of Art" in the series.

openings in the wall of the circular engine-house, upon which are fixed the bearings on which they oscillate; the other end of each beam is connected with one of the pumps. The steam-cylinders consist each of two separate concentric cylinders, of which the outer one, consequently a ring-cylinder, is 12 feet diameter, the inner one being of 7 feet diameter. The steam enters on the under-side of the inner piston only, at a high pressure, by which the cross-head, outer piston and inner end of the pump beams are raised. Arrived at the top of the stroke, the *cataract* comes into operation, and causes a certain amount of pause, after which the partially expanded steam from the bottom of the inner cylinder enters at the top of both this one and the outer cylinder, and in expanding brings down the cross-head and pump beams, while the power is increased by a vacuum being formed under the piston of the outer cylinder.

163. Before entering into a description of some horizontal pumping-engines that have been recently carried out, we will now shortly consider the construction of the component parts of a pump.

1°. The *suction-pipe* serves to conduct the water from the well to the pump-barrel; its diameter should not be less than that of the delivery main, while its vertical height, measured from the surface of the water in the well to the suction-valve, must be rather less than 32 feet, *i.e.* the column of water balanced by the pressure of the atmosphere, though it may be of considerable horizontal length. If $v =$ the velocity of the water in the suction and delivering mains, in feet per minute, $q =$ the quantity of water in cubic feet to be delivered per minute, $d =$ the diameter of suction and delivery main, then

$$d = \sqrt{\cdot 7854 \frac{Q}{v}}$$

The velocity of the water should average from 3 to 4 feet per second.

2°. *The pump-barrel* or *cylinder* should, when small, be

made of gun-metal, and when of larger dimensions and worked with an ordinary piston, should receive a gun-metal *lining*. Where it has a *plunger-piston*, however, this is not necessary, as the plunger generally only fits the cylinder at the top where it is made watertight, as shown in Fig. 10, Plate XVI., where *a a* are two leather rings, *b* a metal one, and *c* the gland. The same arrangement of packing may also be used for ordinary cylinder glands.

The proportion of the diameter of pump-barrel to the length of stroke depends on various circumstances arising out of the peculiarities of individual cases; in general, where possible, we should prefer having a large diameter with a small stroke, as this would give a small speed for the piston (and consequently less wear and tear) with a considerable velocity of the water in the pipes. Where the quantity of water to be raised per minute and the length of stroke are given, the theoretical diameter, *d*, would be found from the equation

$$d^2 \text{ ft.} \times \cdot 7854 \times l \text{ ft.} \times n = q$$

where $l =$ the length of stroke, $n =$ the number of strokes per minute, and $q =$ the quantity of water per minute in cubic feet. But in practice we find that we must add a little to this to allow for inexact workmanship, and we have

$$d = \sqrt{\frac{Q \times 1 \cdot 1}{\cdot 7854 \times l \times n}}$$

3°. The *bucket*, or *piston*, obtains various methods of construction, according to the work for which the pump is intended. The most common construction of *valve-bucket* is that with so-called *butterfly*, or *flap*, valves, which are generally simply leather flaps opening up against the pump-rod, having a leaden or brass disc fixed at top and bottom. A very superior valve-piston is now much employed, particularly for the *air-pumps* of steam-engines, as shown in Fig. 11, Plate XVI. The body of the bucket *a*, is of brass, and is formed with a grid on the top, as shown in plan; the ring *b*, is adjustable by means of the screws *e*, for the purpose of pressing down the packing, which is of hemp; *c* is a disc of vulcanised india-rubber, which rests on top of the grid, and is held in such a manner that it can slip round easily, thus preventing the same place from always striking on the ribs of the grid; *d* is a gun metal guard, against which the india-rubber disc flaps when the water is passing through the bucket; this guard must be furnished with a number of small holes to prevent the india-rubber from adhering to it by suction. The india-rubber is strengthened by being formed with a layer of canvas in it.

Another form of valve-bucket is shown in Fig. 12, where *a* is the body of the bucket, *b* a ring for pressing down the packing, which is done by means of the fork fixed to it being pressed down by the nut *d*, upon the bucket-rod; *c* is the valve which slides upon the bucket rod and is lifted by the water against the stop *e*. Fig. 13 represents a simple *solid* piston, where *a a* are metal discs, and *b b* leathern ones; these, even when worn, will be perfectly water-tight, on account of the water pressing them against the sides of the cylinder. In Fig. 14 is shown the construction of a larger description of solid piston, these are sometimes also made with gun-metal rings for packing, as for those of steam-cylinders. The *plunger*-piston is formed solid when small, and when of a large diameter is cast cylindrical, closed at the bottom; it is often connected directly with a crank, in which case the piston-rod at the same time forms the connecting rod, and is hinged at the bottom of the piston. These pistons are generally used for pumps of small dimensions, particularly for feed pumps.

4º. *The valves.* These are a most important feature in the construction of pumps, as the efficacy of the same depends in a great measure on the manner in which the valves are constructed. As they are always liable to get out of order, no matter how well constructed, they should always be so arranged that they are easily accessible, this is generally done by having an opening close to their seat, which is closed by a cover being bolted on. Another important point is, that the area through which they allow the water to pass should always be as large as possible, approximating to the area of the suction or delivery pipes, and that at the place where the valves are fixed, the diameter of the pipes should be enlarged to such an extent that the water has plenty of room to pass away from them. The valve generally consists of two parts: the actual *valve* and its *seat*, both of which should be of gun metal; the *seat* is fixed firmly into the *valve-chamber*, which is usually made a separate casting.

The most common form of valve is what is termed the *conical valve*, a construction of which is shown in Fig. 15, Plate XVI., where *a* is the *valve-seat*, made tapering so as to fit firmly into the valve chamber. Upon the rim of this is fixed the bridge *b*, serving both as a guide and stop for the valve; the bush *d*, is cast in one piece with the seat being held by three or more ribs. The valve itself, *c*, is dished, and where it rests on its seat is turned conical; to ensure perfect fitting, the valve and seat are generally ground together before fixing. It is cast with the spindle *e*, which passes through the bridge *b*, and bush *d*, and has a shoulder formed upon it to prevent the valve rising too high. For valves of a large description, *double-beat* valves are generally employed, of which the ordinary construction is shown in Fig. 8. This is so arranged that the valve on opening presents two outlets for the water, one at *a* the other at *b*; in closing, the valve *c* drops upon the gun-metal rings *d*, *e*, fixed in the seat *f*, which is of cast-iron; this is cast with a cylindrical portion *g*, which serves as guide to the valve, as do also the ribs *h*, *h*. We shall describe other forms of valves when speaking of the pumps to which they have been applied.

It is essential, for the preservation of valves in good working order, that the opening, and more particularly the closing of the same, should occur without any violent concussions; the manner of effecting this consists in making the valve of such a weight that it is just balanced by the pressure of the water when passing through it. By this means, on opening, instead of being thrown with great force against the stop by the flow of water, it rides upon the current, in a similar manner (to illustrate by a familiar phenomenon) that the pea dances upon the current of air blown through the stem of a tobacco pipe. On closing, the valve gradually sinks as the pressure of the water diminishes under the control of the crank, and at the end of the stroke it is so close to its seat that it shuts without any perceptible concussion. This highly judicious arrangement only involves a very slight increase of head to compensate for the pressure of the valve, and this may easily be calculated in the following manner:—

We have first to ascertain the weight of the valve; then, by dividing this by the area in square inches which it presents to the action of the water, we find what pressure per square inch of that surface must be exerted by the water flowing through, in order for it just to balance the valve; this pressure, divided by the weight of a cubic inch of water, will give the column of water, in inches necessary to lift the valve. Again, if we divide the volume of water in cubic inches, which the pump delivers per second, by the area of water-way in the valve-seat, we obtain the velocity of the water when passing through the valve-seat, and this velocity must be, as near as practicable, equal to that resulting from the column of water due to the weight of the valve. We shall illustrate this by a numerical example, in the following description of pumping-engines.

164. In calculating the power required for working pumps, we have not only to take into consideration the actual height to which the water has to be raised, but also a certain additional resistance resulting from the friction of the water against the pipes. If, namely, we have a certain length of pipe laid perfectly level, we shall find that in order

that the water should flow through the same with a given velocity, it will be necessary to exercise a certain pressure, represented by a *head of water* variable according to the diameter of the pipe and the velocity required. For further information on this subject we must refer the reader to the "Rudimentary Treatise on Civil Engineering," Parts III. and IV. (in this series), or to other works on hydraulic engineering.

SECTION II.

165. *The Crystal Palace Pumping-Engines.*—These pumping-engines, of which there are two pair, were designed by Mr. E. A. Cowper. One pair, of which one is shown at Fig. 1, Plate XVI., serve to raise the water from the lower reservoir or lake to the upper open reservoir in the grounds of the Crystal Palace, having a head of about 103 feet, and delivering 2827 gallons per minute, with which they supply the intermediate fountains; they are fixed close to the lower reservoir, and have consequently only a short length of suction-pipe.

The other pair, shown in plan and transverse section in Figs. 3 and 4, are of the same construction as the first, differing only a little in dimensions; their duty is to supply the before-mentioned lower reservoir. In this case, in order that the engines should be out of sight, it was found necessary to place them at a considerable distance from the lake from which they draw, the suction-pipe thus obtaining a horizontal length of about 900 feet. To obviate the shock that would result from the fluctuation in the velocity of the water due to the stroke of the pumps, a large *stand-pipe* was fixed on the suction-pipe close to the pumps. This stand-pipe as shown at Y, Figs. 3 and 4, is a rectangular cast-iron tank, 7 feet 4 inches long by 2 feet 4 inches wide, and 22 feet high, into which the two suction-pipes V V enter, after communicating with the pumps by means of the vertical suctions V′.

166. In the upper pumping-engines the suction-pipes take the water from the well immediately beneath the pump-cylinders, as shown in Fig. 1. Of these engines, the steam-cylinder A, is 35½ inches diameter, by 36 inches stroke, while the pumps B, are the same stroke and 21½ inches diameter; these are double-acting, and are placed horizontally, being fixed opposite the steam-cylinders, with the cross-heads D of whose piston-rods the bucket-rods are also connected. In order to render these pumps, which would otherwise be *direct acting*, controllable by the crank-motion, the cross-head D, which slides in the guides E, fixed to the frame F, is attached to the ends of a forked connecting-rod G, which passes behind the pump-cylinder, and is there connected with the crank upon the fly-wheel shaft H, by which the two engines are coupled at right angles. The fly-wheel I is 14 feet diameter, and runs between the two engine-frames F. Each of these frames is cast in one piece, and a girder passing underneath them connects them with the fly-wheel-shaft, the plummer-blocks of which are fixed to it. By this means the strains in the engines are all self-contained, thereby obviating the massive foundations required in an ordinary beam-engine. The air-pump K of the engine is placed vertically below, and is worked by bell-crank levers L, connected by links to the cross-head D, the stroke being half that of the steam-piston. The air-pump is 22 inches diameter, its bucket and delivery valves being india-rubber discs upon brass grids, as described above; the foot-valve is here dispensed with, as the bucket dips into the well sunk in the bottom of the condensor N. The injection is admitted by the conical valve O, the opening of which is regulated by a screw; with this form of valve the condensation is rendered very perfect as the water enters with considerable force in a thin film, which, in striking against the sides of the condensor is thoroughly distributed throughout the chamber, and thus produces a vacuum of 13·6 lbs. per square inch. The cold-water pump is worked

by the other end of the bell-crank levers L, its valves being likewise india-rubber discs on brass grids.

The pump-cylinders B, are of cast-iron, turned inside, and fitted with a gun-metal lining $\frac{3}{8}$ inch thick; to these are attached, below, the chambers of the suction-valves R, and above, those of the delivery valves S, on the top of which are fixed the cast-iron air-vessels T, of cylindrical form with hemispherical ends. These air-vessels, which are 3 feet 9 inches diameter, and 8 feet 3 inches in length, are provided with glass gauges U, to indicate the height at which the water stands in them. Any leakage of air or insufficient volume of the same is made good by fixing a small air-cock upon each suction-pipe, by which a small quantity of air is drawn into the pump and passes to the air-vessel at each stroke. From the air-vessel of each pump the water passes through the delivery mains W, which join into one single main X, in the centre between the two engines, as shown in the transverse section, Fig. 4. Each of the delivery mains W, are provided with a *stop-back valve* which keeps the pressure in the mains from the pumps when they are not working, or are under repair.

The construction of the valves is shown in an enlarged section and plan, Fig. 7; they are what are termed *ring-valves,* designed by Mr. Cowper, and are of great efficacy, owing to the large area of *water-way* that they present, with a very small amount of lift. Both the valve *a*, and seat *b*, are of gun-metal, the valve consisting of two rings *c* and *d*, fitting on to two corresponding annular openings in the seat, both an inch wide, of which the lower one is 7 inches mean diameter, and the upper one 13 inches, giving a total area of 63 square inches for the passage of the water, and an equal area to this is given by the four cylindrical openings resulting from a lift of $\frac{9}{16}$ inch of the valve. The two rings *c* and *d*, are connected together by means of the ribs *a*, and by these to the gland, which slides upon the centre pin *e*.

The valve weighs 37 lbs., which, divided by the 63 square

inches lifting area, gives a pressure of ·59 lbs. per square inch, necessary for lifting the valve; this pressure corresponds with that which would be exerted by a column of water of 16·1 inches in height, giving a corresponding velocity of the water of 9·25 feet per second, and this velocity is somewhere between the actual maximum and mean velocities of the water through the valve-seat, when the pumps are working at their ordinary rate of 15 revolutions, or 30 strokes, per minute. The valves are thus just balanced by the flow of the water, and in consequence of this, and of the gradual increase and decrease of velocity due to the crank motion, they are made to open and shut without any perceptible concussion whatever. The valve-seats are fixed to the valve-chamber by means of the cross-bars *f*, which fit into pockets on the sides of the chamber, and receive the thrust of the set screws *g*.

The steam-cylinders work expansively by means of the cam and rocking levers fixed upon the crank shaft H, which actuate the slide-valves at the sides of the steam-cylinders. The steam enters at a pressure of 18 lbs. per square inch above the pressure of the atmosphere, is cut off at $\frac{1}{3}$ of the stroke, and expands down to 5·8 lbs. below the atmosphere, as shown in the indicator diagram in Plate I., Fig. 4, giving a mean pressure of 5·76lbs. above the atmosphere; the mean vacuum is 12·44 lbs. below the atmosphere, and the total mean pressure is therefore 18·2 lbs. per square inch. This gives 49·1, indicated H. P., for the pair of engines, at 15 revolutions per minute, or 89 per cent. useful effect.

SECTION III.

167. *Yarmouth Water-works Pumping-Engines.*—These pumping-engines, of which there are two coupled at right angles, also designed by Mr. E. A. Cowper, are employed for forcing the water to a high-level reservoir, at a distance of about four miles. Their construction is

similar to those just described, having each, in addition, a low-lift pump, situated beneath the engines, which serves to pump the water from a collecting reservoir on to a set of filter beds.

Fig. 2, Plate XVI., represents a longitudinal section through one of these engines, and Fig. 5, a plan of both to a smaller scale. The steam-cylinders A, are 24 inches diameter, with 36 inches stroke; the high-lift pumps, B, worked directly from the cross-head D, of the steam-cylinder piston, are 9 inches diameter, with 36 inches stroke. A forked connecting rod G, is carried from the cross-head D, to the cranks H, upon the fly-wheel shaft; the fly-wheel I, which is in the centre between the two pumps, is 12 feet diameter. These engines are placed over a well, as shown on plan; and the frames F, are carried across the same by means of girders underneath: of these, the inner ones are continued to carry the plummer-blocks for the crank-shaft. The high-lift pumps B, are supplied with filtered water by a main V, from the filter beds, and as there is some head of water upon them, a stand-pipe Y, is arranged on the end of the main V, to allow of a slight fluctuation of the level of the water, due to the small variation in the rate at which the pumps take the water. The actual height to which these pumps have to lift the water is about 100 feet, to which, however, has to be added a further height of 60 feet, due to the friction of the water in the long length of main, making a total height of 160 feet, to which the pair of pumps have to raise 800 gallons of water per minute.

The low-lift pumps P, are single-acting plunger-pumps, $18\frac{3}{4}$ inches diameter, with 18 inches stroke; they are worked by means of the bell-crank levers L, that work also the air-pumps K, of the engines; the plunger is loaded to such an extent as to cause half the work to be done during the up-stroke of the same. The height to which they have to raise the water is only 20 feet, at the rate of 1200 gallons per minute the pair of pumps; on account of this small amount of head, the low-lift pumps work without an air-vessel. The high-lift pumps have one large air-vessel T, between the two engines, in the place of one to each pump, as in the former case: and the suction-valves R, are in chambers at the side of the pumps, instead of underneath them, as seen in the plan, Fig. 5. The delivery-valves S, communicate by means of the pipe W, which, before entering the air-vessel, is provided with a stop-back valve at U, for purposes before explained. The valves to the high-lift pumps are shown in an enlarged section and part plan, Fig. 6; they are *double-beat valves*, differing, however, in construction from the ordinary ones shown in Fig. 8, inasmuch as the larger area of opening is at top instead of at bottom, and the valve *a*, which is of gun-metal, slides upon the centre pin *c*, similar to the ring-valves just described. This arrangement is preferable to that shown at Fig. 8, on account of the much greater length of the gland *d*, in proportion to its diameter, giving less chance of the valves canting and sticking, which sometimes occurs with those as at Fig. 8. The seat of the valve *b*, is of cast-iron, with gun-metal facings where the valve rests. The openings to these valves are respectively 5 inches and 8¼ inches internal diameter, giving an area of water-way of 26 square inches, with a lift of $\frac{11}{16}$ inches for the valve. The pressure required to balance the weight of these valves, which is 18 lbs. is 68 lbs. per square inch, equal to a column of water of 19 inches, which corresponds to a velocity of about 10 feet per second. This is about the velocity of the water when passing through the valve, when the pumps are going at the ordinary rate of 30 revolutions, or 60 strokes, per minute. The valves for the low-lift pumps are of the same construction, being somewhat larger than those of the high-lift pumps; they have a water-way of 88 square inches, and weigh 42 lbs., requiring a pressure of 5 lbs. per square inch to lift them, which is equal to a 14-inch column of water, or a velocity for the water of about 8 feet 8 inches per second.

The air-pumps K, and condensor N, are of the same construction as those for the Crystal Palace pumping-engines; the cold-water pumps are here rendered unnecessary by the condensor being immersed in the pump-well. The slide-valves, of which there are two, one working over the other, for effecting the expansion of the steam, are worked from two eccentrics on the crank-shaft. The steam, which enters the cylinder at a pressure of 9 lbs. per square inch above the atmosphere, is cut off at $\frac{1}{3}$ of the stroke, and expands down to 7lbs. below the atmosphere, giving a mean pressure of 1·05 lbs. above the atmosphere; the mean vacuum is 13·17 lbs. below the atmosphere, giving therefore a total mean pressure of 14·22 lbs., equal to 70 indicated H. P., for the pair of engines, at the rate of 30 revolutions per minute. This gives 81 per cent. useful effect, exclusive of the friction of the pumps.

The low-lift pumps, as also the air-pumps and condensor, are connected in a rigid manner to the engine-frames by means of the framework Z, connected at top to the underside of the girders carrying the engines, and below to the upper-side of the girders, carrying the low-lift pumps, &c.

168. Besides the advantage of equalising the flow of water in the mains, the arrangement of having a pair of pumps coupled at right angles, in place of one to do the same duty, presents the further convenience of being able to work one engine while the other may be undergoing repair; or of working one engine instead of both, when a less quantity of water is required.

We shall describe another kind of horizontal pumping-engines in the next chapter, they forming a part of the machinery to be considered therein.

CHAPTER IX.

HYDRAULIC CRANES.

169. We have elsewhere considered the beneficial action of the fly-wheel as an accumulator of power, in enabling a *motor*, of comparatively small momentary intensity, to overcome, at certain intervals, resistances surpassing many times the actual power of the motor. This principle may be adopted from economical motives in all such cases where, at certain intervals of time, great dynamical effects are required, which would otherwise necessitate the great expense of erecting engines of large proportions that, perhaps, during the greater part of the time, would be standing idle.

Sir William Armstrong has adapted this principle with very great success to the working of cranes and various other machines, by accumulating the power required at certain intervals by them in the form of a constantly loaded column of water, which is maintained in a receptacle, termed *accumulator*, through the agency of a small pumping engine always at work. From this *accumulator* the power is transmitted to the cranes, &c., on the well-known principle of hydrostatic pressure, through the medium of pipes of a small diameter, the water under pressure being brought to act in certain cylinders provided with pistons, the motion of which causes either the raising and lowering of weights by means of the crane chain, or else the swinging round of the jib. These constructions are so admirably arranged that, with the aid of one single man, the greatest loads may be raised or lowered, or swung round by the crane with a rapidity and ease of motion truly marvellous.

This hydraulic apparatus has also been employed very advantageously for opening and shutting large dock-gates and swing-bridges, for working hoists and turntables, and various other machines requiring occasionally motive power in a concentrated form. It is now employed for such purposes at the West India, London, Victoria, and St. Katherine Docks, as also at the Woolwich Arsenal, Great Western Railway Terminus, and at all the most important ports of the kingdom.

170. In the first applications that Sir William Armstrong made of this principle, he employed a *natural* head of water as the motive agent, obtaining the same by pumping water into tanks at an elevation of about 200 feet; but subsequently he has always employed the *accumulator*, as offering the advantages of greatly increased capacity for pressure, and a less prime cost of erection. The accumulator is shown in Fig. 3, Plate XVII.: it consists of the large cast-iron cylinder *a*, fitted with the plunger *b*, which works water-tight by means of the gland *c*, and packing. To this plunger is attached, by means of the bolts *f*, and strong cast-iron cross-head *e*, the loaded weight-case *d*. Thus a pressure is obtained upon the water in the cylinder, equal to a column of water 1500 feet high, or 660 lbs. upon the square inch. As the water is pumped into the cylinder by the pumping engines through the pipe *h*, the piston, with the weighted case rises, being guided by the strong wooden framework *g*, and is made to regulate the amount of water pumped in, by actuating a throttle-valve in the steam-pipe of the pumping engine, which it closes after having reached a certain height. When the cranes, &c. are in operation, the water passes from this cylinder through the pipe *i*, to those actuating the motion of the cranes, and the weighted plunger naturally descends, always keeping up a constant pressure upon the water; in descending, the same causes the throttle-valve to

open again, and the water is again pumped in. This accumulator is made of such a size as to be able to supply all the cranes, &c., when working simultaneously; but when the hydraulic machinery extends over a consideraahle range of ground, it is found advisable to have an accumulator at each end to prevent having too great a body of water in motion in the pipes. The loss of pressure due to the friction of the water in the pipes is very slight, owing to the small velocity of the water and the great head upon the same.

171. The pumping engine by which these accumulators are charged is shown in section in Fig. 9, Plate XVI.; the pumps, which are horizontal, are connected directly with the piston-rod of the steam-cylinder, but their action is controlled by that of a crank and fly-wheel. *a* is the horizontal steam-cylinder, the piston-rod of which extends both ways, and is connected at either end with the single-acting plunger pumps, *b*. To the cross-head, *c*, which works in guides, is attached the connecting-rod *d*, imparting motion to the crank *e*, and fly-wheel *f*; the other end of the piston-rod actuates the slide-valve *g*, by means of the lever *h*, and link *i*. *j j* are the suction, and *k k* the delivery valves, which are conical, and weighted; *l* is the suction-pipe, and *m* the delivery main in connection with the accumulator. The suction mains take the water usually from a tank above the engine house, into which the water that has been spent by the cranes is discharged, so that the same water remains continually in use and is thus kept free from foreign matter likely to injure the action of the machinery.

172. In Figs. 1 and 2, Plate XVII. are represented an elevation and plan of the most modern construction of a *hydraulic crane* capable of raising 1 ton.

Both the *jib a*, and the *crane-post b*, are of wrought iron; the latter is stationary, being fixed into the masoury, and also held by the plate *f*. The jib is connected with the tie-rods by means of the cheeks *d*, which carry a cross-head *d'* at the top, swinging on a swivel attached to the crane-post; upon this cross-head is also fixed the guide pulley *e*, for the chain. At the bottom the jib is attached to the casting d'', in which the roller d''', serving to transmit the thrust of the jib on to the crane post, is fixed. This casting is also provided at *h* with a circular groove for receiving the chain which actuates the swinging round of the jib in the manner to be described hereafter.

The crane chain which is provided, at *i*, with a weight to lower the same, passes over the guide pullies j, j', e, and down the centre of the crane post, from which it passes over the guide pulley *k*, and is then connected with the lifting apparatus now to be described.

l is the pipe in communication with the accumulator; upon this is fixed the *lifting* or pressure valve *m*, by means of which the water under pressure is admitted to, or shut off from, the lifting cylinder *n*, by passing through the pipe n'. Another valve *o*, connects this pipe n', with the exhaust-pipe o'. These two valves are kept closed by means of the loaded levers *p* and *q*, and are both worked by the lever *r*, which when pulled over in the direction of. the valves, opens the *lift valve m*, and when pulled the other way opens the exhaust valve *o*. The degree to which these are opened regulates the speed with which the *lifting ram s*, moves out of or into the cylinder *n*. This cylinder is fixed to the foundations by means of the brackets n''; and it must be observed that these foundations, as far as they are required for the lifting apparatus, need only be of a very trifling nature, as the strains exerted by the same are almost entirely self-contained.

The water being admitted into the lifting cylinder, it presses upon the end surface of the ram with the same pressure per square inch as is being exerted upon it by the loaded plunger in the accumulator. This pressure causes the ram to move out of the cylinder and thus impart motion to the crane chain, which is first fixed to the cylinder at *t* (as shown by the dotted line), then passes successively over one of the pullies *u*, in the sheave attached to the end of the lifting-ram, and one of those fixed at *v* to the end of the cylinder; the chain thus travels backwards and forwards and at length passes over the guide pulley *k*, attached to the crane-post and up to the top of the jib in the manner described. By this arrangement, the extent of motion in the lifting ram is increased in the chain at *i* as manifoldly as it passes up and down over the pullies *u* and *v*; that is, in this case six times, as it is evident that, if the ram moves out to a distance of one foot, then the chain at 1 is also increased by a length of one foot, which necessitates a motion of one foot in 2; but 2 is itself increased by a length of one foot due to the motion of the ram, its actual motion is therefore two feet; in the same manner 3 moves three feet, &c., and lastly, 6 six feet. The ram, during its motion out or in, is steadied by the guides *w*, and in order that it should not come with too great a force against the stops forged upon these guides at the end of the stroke, when nearly at that point a tappet, fixed to the cross-head at *u'*, comes in contact with a projection upon the sliding-bar *x*, and causes the latter to move forward, thus actuating the closing of the lift-valve *m*.

When a load is on the crane-chain, the same is lowered simply by opening the exhaust-valve *o*, when the lifting-ram would be forced back into the cylinder by the pull on the chain only; but as the chain also requires to be lowered without any load upon it, and as it would not be advisable to make the bob at *i* of too great a weight, a *return cylinder y* is employed, the ram of which is connected with the cross-head of the lifting ram, and the pressure exerted in it serves to force back the latter. The pressure upon the ram of the return-cylinder is constant, and we have therefore, in calculating the actual power of the lifting-ram, to deduct the area of the former from that of the latter. To operate the swinging or *slewing round* of the jib, the two small hydraulic cylinders A A, fixed to the framework *f*, are brought into action; the rams B, of these are furnished with the pullies C, over which a chain, attached to the cylinders at D, passes, and thence round the groove *h*, on the casting *d''* connected with the jib. By this arrangement, when one ram is forced out by the water pressure, the other one goes in, and the jib is swung round in one direction; when the other ram is forced out, the jib is moved in the reverse direction. The power is supplied to these rams by means of the pipes E, which communicate with the supply pipe H, and the exhaust pipe G, by means of a three pool valve at F, worked by a lever at I; this valve connects simultaneously one cylinder with the supply, and the other with the exhaust. As in slewing round the jib, the same acquires some momentum, an objectionable shock would take place on suddenly closing the valves; to obviate this, *relief valves* are fixed upon the supply and exhaust pipes, which open *against* the ordinary pressure and communicate with the cylinders. By means of these, when the ordinary valves are closed and the momentum of the jib endeavours still to give motion to the rams, this is permitted to a certain extent by the water in the cylinder, of which the ram is moving *in*, forcing the relief valve on the supply-pipe open and injecting the water into the same, while the ram which is moving out draws water through the relief valve upon the exhaust-pipe, to supply the vacuum formed. A small cistern M, communicates with the exhaust-pipe G, by means of which the water is more readily supplied for this purpose. As is the case with the lifting ram, the turning rams are capable, at the end of their stroke, of closing the three pool valve F, by the cross-head of one ram striking against the tappets J J, upon the sliding bar K, and thus moving the latter either forward or backward. By the arrangement of the chain, the extent of motion in the same is double that of the rams. These are steadied in their motion by the guide-bars L L.

173. In cases where a greater extent of motion is required than can be afforded by the arrangement of cylinders and pullies as above described, Sir William Armstrong has communicated a rotatory motion to a drum-shaft by means of three cylinders with plunger pistons, which actuate three cranks upon the drum-shaft placed consecutively at one-third of the circumference. The valves to these cylinders, which consist of conical weighted spindles, are worked by means of revolving cams upon the same shaft. Where a natural head of water of about 250 to 300 feet elevation occurs, he has also produced a rotatory motion by means of cylinders and pistons similar to those in an ordinary high pressure steam engine, which were worked by means of slide valves, having also relief valves to prevent detrimental shocks at the end of the stroke.

It will be self-evident that these hydraulic machines may be employed to advantage even where the accumulators are charged simply by manual labour by means of ordinary force-pumps, such as are employed for working hydraulic presses, as a less number of men will be required by the continual working of these pumps than would be, to produce the same effect in a concentrated form and in a direct manner upon the hoisting machinery; and owing to the intermittent nature of the work, the men would, in the latter case, be continually standing idle.

THE END.

BRADBURY AND EVANS, PRINTERS, WHITEFRIARS.

Printed in Dunstable, United Kingdom

91221347R00117